M. A. Stein

Detailed Report of an Archaeological Tour

With the Buner Field Force

M. A. Stein

Detailed Report of an Archaeological Tour
With the Buner Field Force

ISBN/EAN: 9783348031608

Printed in Europe, USA, Canada, Australia, Japan

Cover: Foto ©berggeist007 / pixelio.de

More available books at **www.hansebooks.com**

PRESENTED

BY ORDER OF

THE GOVERNMENT OF THE PUNJAB.

ARCHÆOLOGICAL TOUR

WITH THE

BUNER FIELD FORCE.

BY

M. A. STEIN, ᴘʜ. ᴅ.,
ᴘʀɪɴᴄɪᴘᴀʟ, Oʀɪᴇɴᴛᴀʟ Cᴏʟʟᴇɢᴇ, Lᴀʜᴏʀᴇ.

LAHORE:
ᴘʀɪɴᴛᴇᴅ ᴀᴛ ᴛʜᴇ ᴘᴜɴᴊᴀʙ ɢᴏᴠᴇʀɴᴍᴇɴᴛ ᴘʀᴇss.
1898.

CONTENTS.

DETAILED REPORT

OF AN

ARCHÆOLOGICAL TOUR

WITH THE

BUNER FIELD FORCE.

———•◆•◦——

I.—PERSONAL NARRATIVE.

AT the end of November, 1897, Major H. A. DEANE, C.S.I.,
Political Agent, Swat, Dir and Chitral, had been kind enough to
call my attention to the opportunity which the punitive expedition,
then under consideration against the tribes of Bunér would offer
for the examination of the antiquarian remains of that territory.
Bunér, as that portion of the ancient *Udyána* which had hitherto
been wholly inaccessible, and as the place from which a number of
Major Deane's puzzling inscriptions in unknown characters had
been obtained, could reasonably be expected to furnish an interesting new field for archæological exploration. I was hence eager to
avail myself of the occasion.

Thanks largely to Major Deane's recommendation and the kind
interest shown in the matter by the Hon'ble Mr. DANE, Chief Secretary to the Punjab Government, and my friend Mr. MAYNARD, the
Junior Secretary, my application to be deputed with the Malakand
Field Force during its operations in Bunér was readily approved

of by the Hon'ble Sir MACKWORTH YOUNG, K.C.S.I., Lieutenant-Governor of the Punjab. The Local Government agreed to bear the expenses connected with my deputation. On the 29th December, when returning from a short archæological Christmas tour in the Swat Valley, I received at Hoti Mardán telegraphic intimation that the Government of India in the Foreign Department had sanctioned the proposal. In accordance with the instructions conveyed to me I saw on the same day at Kunda Camp Major-General SIR BINDON BLOOD, K.C.B., Commanding the Malakand Field Force, who very kindly assured me of his assistance in connection with the proposed archæological survey. He also informed me of the early date fixed for the commencement of the operations against Bunér. I had just time enough to hurry back to Lahore, where the Annual Convocation of the University required my presence, and to complete there the arrangements for my camp outfit and for a Surveyor from the Public Works Department who was to accompany me.

On the afternoon of the 4th January 1898 I left Lahore after assisting at the Convocation held under the presidency of the Hon'ble the Lieutenant-Governor and Chancellor of the University. Starting from Nowshera Station on the following morning I caught up on the same day General Blood's Division while encamped at Katlang on its march towards the Bunér border. Heavy rain on the preceding day had made the air remarkably clear. As I passed through the breadth of the great valley which forms the ancient Gandhára, the barren mountain ranges enclosing it on the north and south stood out with a boldness reminding me of classical regions. From Mardán to Katlang the rugged Pajja Range, which in its secluded straths and nooks hides a number of ancient sites, kept all the way prominently in front. On a small spur descending from this range, which is passed to the east of the road close to the village of *Jamálgarhi*, the ruins of the large Buddhist monastery came into view, which was excavated here by General Cunningham. I was unable to revisit these interesting remains for want of time, but was informed that numerous injured torsos of statues which had been brought to light by those diggings, still cover the ground in several of the Vihára Courts.

At Katlang I was joined by Fazl Iláhi, Draftsman, from the office of the Executive Engineer, Pesháwar, who was to act as my Surveyor. There I found also Sherbáz, Jamadár of Swat Levies, and Kator Shah, a Mián from Sháhbázgarhi, whom Major Deane had kindly sent to accompany me to Bunér and to assist me by their local knowledge.

On the 8th January the force moved from Katlang to *Sanghau*, at the entrance of the defile leading to the *Tangé* Pass which had been selected as the route for the advance into Bunér. A reconnaissance conducted by General Blood up the defile showed that the pass was held by a gathering of tribesmen under numerous standards. Accompanying this reconnaissance, I came in the narrow ravine through which the path leads, and about a mile and a half above Sanghau village, upon unmistakeable traces of an ancient road. I was able to examine these before the Sappers had commenced their work of improving the track. In several places where the present path runs along rocky cliffs high above the stream draining the gorge, I noticed supporting walls of rough but solid masonry. They resembled closely in their construction the walls over which the ancient so-called " Buddhist" roads on the Malakand and Sháhkót Passes are carried in parts. Higher up in the defile the traces of this old road seem to be lost. At least I did not come across any on the following day either on the track chosen for the transport route or during my climb up the hillside to the north.

When returning to the camp it was too late to examine closely the ruins which were pointed out to me as those of 'old Sanghau' on a spur about 1⅓ miles to the east of the village. Seen from below they appeared to consist of groups of solidly built old dwelling-places, such as are found in great numbers covering the hillsides at various points of the Lower Swat Valley. About half a mile

Sanghau.

further in a north-easterly direction old remains are said to
exist near a large spring, the water of which is now brought by a
stone-conduit down to Sanghau village. A great deal of ancient
Buddhist sculpture has been extracted at various times from
ruined sites near Sanghau, but it is only of the excavations conducted for General Cunningham that some account can be traced.

The night passed in camp at Sanghau, and thus yet within
British territory, brought some "sniping" which was attributed
by competent judges to 'loyal' subjects of that neighbourhood.
On the afternoon of the following day the Tangé Pass was taken
after a prolonged artillery fire and some fighting. While the
Patháns, Sikhs and Dográs of the XXth Regiment, Punjab Infantry,
climbed in splendid style the high peak commanding the pass on
the west, the Highland Light Infantry and West Kent Regiments
carried the naturally strong position of the enemy in front. I watched
the interesting engagement from the spur occupied by the
mountain batteries in action and climbed up to the narrow rocky
ridge which forms the pass, soon after it had been taken. From
that commanding height, circ. 3,800 feet above the sea, there
opened a wide view over the western portion of Bunér bounded
in the direction of Upper Swat by Mounts Ilm and Dosirri.

Tangé Pass.   At a point where the crest forms a salient angle to the west,
and about 300 yards from the saddle by which the mule-track
crosses the pass, I noticed the remnant of what was probably once
a small fortification, in the form of a semi-circular platform built of
rough masonry. The outside wall supporting it was traceable for a
length of 20 feet. The tribesmen holding the pass had raised one
of their main sangars on this very platform. The gathering of
standards I had noticed near this spot in the early part of the day
showed that it had been considered important and held in force
also by the most recent defenders of the pass. The absence of
other traces of old fortification on the ridge is easily accounted for

by its extreme narrowness and the steepness of the cliffs on its western face. These cliffs themselves would form a sufficiently strong line of defence against any enemy not armed with modern guns. On the Tangé Pass there was thus neither room nor need for such extensive fortifications as can still be traced in ruins of evidently ancient date on the Malakand and Sháhkót Passes.

Accompanying the troops of the 1st Brigade which I still found on the crest of the pass, I reached by nightfall *Kingargalai,* a Bunér village belonging to the Salárzai tribe, situated in the valley some two miles from the eastern foot of the pass. This small village formed our quarters for that night and the next two days. The forcing of the pass had apparently put all thought of open resistance to an end. This and the neighbouring villages were found completely deserted, but Jirgas of the Salárzai and other adjoining tribal sections were soon coming in to treat for terms. General Meiklejohn, Commanding the 1st Brigade, hence kindly allowed me to start already on the morning of the 8th January with a small escort for the inspection of the extensive ruins plainly visible to the west of Kingargalai on the spurs sloping down into the valley.

The most conspicuous groups of ruins were found situated on a series of rocky ridges which jut out, with a general direction from north to south, into the valley leading to the north-west of Kingargalai towards the *Nawedand* Pass. They form the extreme offshoots of spurs descending from the high peak to the west of the pass, which has already been mentioned. The largest of these ridges, which also bears the most prominent of the ruins, lies at a distance of about 1¾ miles from *Kingargalai.* All along the crest of the ridge and also for a short distance down its slopes are found separate groups of ruined buildings. They are erected either where small level shoulders give sufficient space, or on walled-up terraces leaning against the hillside. Their general plan and construction clearly prove them to be the remains of ancient dwelling places. The walls consist of solid masonry resembling closely in its construction that seen in the walls of the

Ruins near Kingargalai.

Takht-i-Báhi Vihâras and other ancient Gandhâra ruins. Large rough slabs, of approximately equal height but irregular shape at the sides, are placed in regular courses ; sufficient space is left between them laterally to allow of the insertion of small flat stones which are placed in little columns, filling the interstices.   Vertically each course of slabs is separated from the next by a narrow band of small flat stones which are put in a single or double row and are intended to adjust slight inequalities in the thickness of the slabs.

This peculiar system of masonry which has been described in the *Archæological Survey Reports*, Volume V, is found in the walls of all ruins of pre-Muhammadan date throughout the territory of the old Gandhâra and Udyâna. It distinguishes them in a very marked fashion from all structures of modern origin which show invariably walls of small uncut stones set in mud plaster without any attempt at regular alignment. Such walls, unless of exceptional thickness, can easily be pulled down with a few strokes of the pick-axe, and when decayed leave after a few years nothing but shapeless heaps of loose stone and earth. The ancient walls on the other hand are of remarkable firmness and have stood the test of time extremely well, particularly where an outer coating of plaster has originally protected them against atmospheric influences. This is sufficiently illustrated by the fact that I have found among the ruined sites of Lower Swat walls of this construction still standing to a height of 30 feet and more. In some instances, too, such walls could be utilized for the foundation of portions of the modern fortifications erected at Malakand and Chakdarra.

The buildings which cover the above described ridges vary considerably in size and plan. Those which occupy sites allowing of greater extension consist of a series of large chambers grouped round a central pile. This is generally raised above the level of the rest by a high base of solid masonry. Plan *I* shows the disposition of a typical structure of this class which stands near the north-eastern extremity of the central ridge above referred to. The interiors of the rooms have been filled up to a great extent by masonry which has fallen from the walls and roofs. The portions of the walls still standing reach in many places only a little above the level of this débris. It is thus impossible to indicate with certainty the position of the doors by which the several apartments must have communicated with each other. In the case of this building the original level of the central rooms marked A, B, C, D seems to have been raised considerably above the ground, as their interior was found now to be nearly 12 feet higher than the rock on which the walls are based. As in the case of similar structures examined in Swat, it is probable that the lower storey of this central pile was built solid; the entrance into the upper storey containing dwelling rooms was through an opening higher up in the wall which could be reached from outside only by means of a ladder. This arrangement, which is clearly designed with a view to defence, is still actually observed in the construction of most village watch-towers across the Afghán border.

That special regard was paid to considerations of safety in the case of most, if not all, the structures here described is evident from the very positions chosen for them. The rocky spurs on which they are found have no other recommendation as building sites except the facilities they offer for defence by their steepness and comparative inaccessibility. The crests of the ridges, which these buildings chiefly occupy, are nowhere less than about 300 feet above the level bottom of the valley. The inconvenience arising from this position in respect of the water-supply, etc., is so great that only an important consideration like that of safety could compensate for it. At the same time it deserves to be noted that these buildings are everywhere standing at such a distance from each other that at a time, when firearms were unknown, none could be said to be commanded by its neighbour. It looks as if the conditions of inter-tribal feud and rivalry which make each man of substance in the average trans-border village watch his neighbour as a possible foe, had already been realized in a far earlier period.

The position which these buildings occupy and the succession of terraces on which some of them rise, give them from a distance more the appearance of small castles than of ordinary dwelling places. They resemble in this respect closely the collections of fortified houses which cover the hill sides at numerous old sites of the Swat Valley like *Landake, Batkhéla, Katgala*, etc. As a distinctive feature, however, it must be mentioned that I have not come across, either among the ruins near Kingargalai or elsewhere in Bunér, the semicircular buttresses which are found very commonly among the Swat ruins at the corners of such structures, in particular of isolated square towers.

To the west of the spur which amongst other ruins bears that shown in Plan *I*, there runs another smaller ridge which with its western scarp faces the side valley of *Manóra*. Along the narrow neck of this ridge too there are numerous ruins of the above description. The ground-plan of one amongst them which represents the simplest type and still shows a well-preserved entrance at some height above the ground, has been reproduced on Plate *II*. On the opposite side of the Manóra Nallah and further up on the hillsides of the main valley towards the Nawedand Pass, I could see other groups of ruined buildings. But the instructions given to me as regards the limits of my explorations on this first day on Bunér soil did not allow me to proceed further in that direction.

Ruins near Tangé Pass. Moving then back to the east along the main hillside, I passed two more spurs running down into the valley nearer to Kingargalai. These were also found to be covered with ruined buildings of the kind already described. Still further to the east at the point where the main valley of Kingargalai is joined by the one leading to the foot of the Tangé Pass, there is a small low spur which has been used as an old building site. At its very end and at a level of only about 50 feet above the flat bottom of the valley, I found the ruin of which a plan is given on Plate *II*. Its peculiar feature is the platform of solid masonry marked *A*, on which rises a small conical mound of rough stones set in layers. The height of the mound is about 11 feet including the base. It appears probable that we have in this mound the remains of a small

Stópa. Unlike other mounds of this character met subsequently during my tour in Bunér it has escaped being dug into by treasure-seekers. Adjoining the base to the east there are two square rooms of which the walls can yet clearly be ·traced. Their construction is exactly the same as that of the walls in the buildings already described. On the floor of the two front rooms (*B*, *C*) there were signs showing that stones and earth had recently been displaced. The Pathán sepoys of my escort, led by an instinct evidently due to experience, at once suspected a hiding place. By removing the topmost stones and then digging down with their bayonets they soon opened two little wells sunk into the ground. They measured each about 5 feet square and were lined with old masonry down to the solid rock. They were found filled with grain and small household property which some neighbouring villagers had evidently deposited there in anticipation of our invasion. There can be little doubt as to these wells having originally been constructed for a similar purpose, small underground store-rooms of this kind having been found under the ruins of the Takht-i-Báhi monastery and elsewhere.

On either side of the short valley running to the foot of the Tangé Pass I noticed several ruined buildings perched high up on isolated cliffs and ridges. They appeared to be similar to those already visited in the valleys towards *Nawedand* and *Manóra*. But the shortness of the remaining daylight made their examination impossible. Considering the number and position of all these ruined habitations, it seems evident that the site to the west of Kingargalai must have been a place of some importance in pre-Muhammadan times. This is easily accounted for by its position on the routes to the Tangé and Nawedand Passes, which both represent important lines of communication. The latter pass in particular, which from all accounts seems comparatively easy for transport animals, opens a very convenient route to the valley of *Básdarra* in the west. From this again the Yusufzai plain to the south as well as the Sháhkót, Chirát and Mora Passes leading into Lower Swat can be reached without difficulty. In this connection I may mention that a coin of *Ooemo Kadphises* (circ. 1st Century B. C.) kindly shown to me by the Chaplain attached to the Highland Light Infantry Regiment was picked up during the occupation of Kingargalai in a small cave on the hill side rising behind the village.

I was unable to ascertain the local name, if any, given by the present inhabitants to the ruins described. The whole population of the valley had fled on the day of the fight on the Tangé Pass, and was still keeping with such cattle as they had managed to save, on the top of the high hill ranges above the valley. It was evident that the occasion, which had thrown Bunér temporarily open, was not the best for collecting local traditions regarding ruined sites from the Pathán inhabitants. Comparatively new-comers to the country themselves and in part migratory as they are, they were often, when got hold of, found unable to give more information than that conveyed by the designation "*Kápir kandare*" ("Káfir ruins"). This is bestowed indiscriminately on all kinds of ancient remains.

Ruins near Nansér.

On the following day, the 9th January, the troops of the 1st Brigade still remained at Kingargalai, while the mule track across the pass was being improved for the transport. I had first hoped to examine the valley further down as far as *Bampókha* which the column marching across the Pirsai Pass was expected to reach that day. But a subsequent order fixed the nearer village of Nansér as the limit of my reconnaissance. This lies about two miles to the east of Kingargalai in a small side valley opening to the south-west. Just opposite to the entrance of the latter the main road of the valley turns round the foot of a very steep and rocky spur which descends from the range to the north. Having noticed high up on this spur walls of ancient look, I climbed up to them and found, at a height of about 500 feet above the valley, two oblong terraces. One is built of solid old masonry along the back of the narrow ridge and extends for about 30 feet from north to south with a breadth of 15 feet. A short distance above, and connected with it by much decayed parallel walls, is a larger walled-up terrace of remarkably massive masonry, placed, as it were, à cheval across the ridge. It measures 45 feet from east to west and 20 from north to south. Its top where nearest to the rocky base still rises to a height of 12 feet above it. There can be little doubt as to this structure having once served the purposes of defence. The position is admirably adapted for this, being approachable only with difficulty over steep cliffs and commanding an extensive view up and down the valley. Small mounds found on the top of these terraces are probably the remains of former superstructures, which being built of less solid materials have decayed long ago. The soil between the rocks on the slopes below is covered with old pottery.

From this point I had noticed villagers descending from the opposite heights to the houses of Nansér, evidently bent on removing property they had left behind on their first flight. As I hoped to receive from them information as to old remains in the neighbourhood, I descended and approached the village. The sight of my small escort was, however, sufficient to cause a fresh stampede of the village folk. When at last after a great deal of parleying some old men were induced to join me, they could only point to a few ruined walls on a hill to the south of the village.

One 'Spingiro' (greybeard), however, knew of a ruined 'gumbas' (dome, circular building) to the west of Kingargalai. As this expression is invariably used by the Pushtu-speaking population of the border for the designation of Stúpas, I did not hesitate to start back under his guidance in the direction indicated. We had passed the ruins examined on the preceding day and proceeded up the Manóra Nallah for nearly two miles further before I could ascertain from my guide that the gumbas he had previously referred to as quite near was in reality beyond the range which forms the watershed towards Bázdarra. To reach the spot and return to camp the same evening was manifestly impracticable at the late hour of day. I was thus reluctantly obliged to turn back to Kingargalai, richer only by an experience of the unreliability of putative distances in the Bunér hills. I had already before heard of the existence of old ruins near Bázdarra, and wish that I may before long have an opportunity to visit that site and other neighbouring localities to the south of the Sháhkót and Mora Passes.

Juvur.   On the 10th I accompanied the march of the greater portion of General Meiklejohn's Brigade to *Juvur*, a large village to the north-east of Kingargalai and below Mount Ilm. The route led for the first four miles down the valley to Bampókha, where the stream which comes from Kingargalai is met by the one flowing from the Pirsai Pass. Before reaching *Bampókha* the road winds round the foot of a detached small ridge which is covered with ruined buildings and terraces resembling those seen near Kingargalai. The short halt made by the troops at Bampókha was not sufficient to allow of an inspection of these remains. A short distance beyond Bampókha the route turns off to the north, and Mount *Ilm* comes prominently into view. This fine peak, 9,250 feet above sea level, with its fir-clad slopes and rocky summit, dominates the landscape in most parts of Western Bunér and forms the boundary of the latter towards Upper Swat. Subsequent enquiry showed that Mount Ilm as the site of more than one Tírtha must have enjoyed a great sanctity in Hindu times. To the west of the mountain is the *Karakar* Pass, the favourite route of communication between Bunér and Swat. In the valley which leads up to the pass lies the village of *Juvur*.

Here the population had not entirely fled, though all houses were appropriated for the accommodation of the troops. I was thus able to collect some information as to old remains in the vicinity. As the Brigade remained at Juvur I could utilize the following day (11th) freely for their inspection. An inscribed stone had been reported to me near the village of *Charrai* situated about two miles to the north-east. But on reaching the spot indicated, which is at the foot of a rocky spur descending from Ilm and about one mile to the north-east of the village, I found that the supposed inscription on a large isolated rock to the right of the path consisted only of a series of cup-shaped holes, probably artificial. The spot is known as *Laka Tiga*.

Returning thence to Charrai, I ascended the narrow gorge, through which the stream of Charrai flows, to an open well-wooded glen known only by the somewhat general designation of *Tangai* (defile; small valley). Tangai which is separated from the Juvur Valley by a low watershed, lies in a direct line about 2½ miles to the north-east of Juvur. Along the slopes of the little spurs, which enclose the glen like an amphitheatre, I found numerous traces of old habitations. Their walls and terraces were generally far more decyed than those of the ruins near Kingargalai. This is in all probability due to the thick jungle which covers this site. The series of fine springs which issue at the foot of the hill slopes and feed the Charrai stream, explains sufficiently the presence of so many ancient dwelling places in this secluded nook of the mountains.

Ascending the spur in the centre of the amphitheatre described, to a height of about 300 feet above the little plain at the bottom of the glen, I reached the rock-cut images of which one of my Juvur informants had told me. The remnants of old walls stretch up close to the foot of the large rock which bears these relievos. The south face of the rock offers a flat and nearly vertical surface about 33 feet long and 30 feet high ; on it a tripartite niche has been cut out to a depth of 3½ inches. It measures 6 feet 9 inches in length and 5 feet in height ; its foot is about 5 feet above the ground. In the centre of the niche is a well-carved relievo figure of Siva, 4 feet 6 inches high, showing the god seated, with his left leg reaching below the seat and the left hand holding the club. On either side of this central image is a smaller figure about 2 feet 9 inches high representing a god seated with crossed legs. The one on the proper left holds in the left hand a lotus on a stalk, and evidently represents Vishnu. The figure on the proper right, which has become more effaced, seems to sit on an open lotus and is probably intended for Brahman. All three figures are surmounted by halos.

Rock sculptures near Juvur.

There can be no doubt as to these sculptures being anterior to the Muhammadan invasion ; probably they are of a considerably earlier date. This may be concluded with good reason from the boldness and good proportions still observable in the design of the relievos, notwithstanding the decay which has overtaken the more exposed portions. To the damage caused by atmospheric influences has been added some chipping done by mischievous hands apparently not so very long ago. Treasure-seekers seem also to have recently been at work here as shown by some small excavations at the foot of the rock. In view of the interest attaching to these sculptures, I regret that no photograph could be obtained of them. They are approached only by a narrow ledge some 3 feet broad, and the rock below them falls off with great steepness. The carvings are thus visible only for one standing immediately before them or from some considerable distance.

The purely Hindu character of these rock sculptures and of those subsequently examined at *Bhai* near *Pádsháh* is a point deserving special notice. It is an additional proof of the fact that Buddhism, which from the exclusive reference made to it in our written records—the accounts of the Chinese pilgrims—may be supposed to have been the predominant creed in the old Udyána, was there as elsewhere in India closely associated with all popular features of the Hindu religious system. This conclusion is fully supported by what other evidence is at present available. Thus the coins struck by the rulers of these regions, from the times of the later Kushans down to the last ' Hindu Sháhiyas, ' show an almost unbroken succession of Hindu, and more particularly Saiva, devices.

Ascending from Tangai to a saddle in the spur to the west, I obtained a good view of the Karakar Pass and the valley leading up to it from Juvur, but did not notice any more ruins in this direction. I then returned to the glen and proceeded to the small rocky hill known as *Níl Dérai*, which flanks the road from Tangai to Juvur on the east. I found it covered on the south face with a series of ancient walls supporting terraces and with masses of débris which evidently belonged to higher structures now completely decayed. These walls stretch up to the very top of the hill which forms a small plateau of irregular shape about 85 yards long from east to west and in the middle about 20 yards broad. All round the top foundations of old walls could be traced, by means of which the available space had been enlarged, and perhaps also fortified. Similar remains are said to exist on the slopes of the higher hill known as *Ghúnd*, which faces Níl Dérai on the western side of the defile leading to Tangai.

On the following day, the 12th January, General Meiklejohn's column marched from Juvur to Tursak by the shortest route which lies in the valley drained by the Charrai stream. As my information did not point to the existence of old remains in this direction I obtained permission and the necessary escort to proceed to Tursak independently by a more circuitous route. This was to enable me to visit the ruins which had been reported to me near *Girárai*, and to see the portion of the main valley of Bunér between Bampókha and Tursak.

Girárai I found to be situated about 5 miles to the south-west of Juvur in a broad open valley which leads to the Girárai and Banjír Passes in the west. About half way I noticed ruins similar in appearance to those of Kingargalai on a detached spur of the hill range to the north of the valley. I could not spare time for their inspection. The locality is known as *Bakhta*. In Girárai itself, which is a village of some sixty houses, the only ancient remain I could trace was a fine ornamented slab built into the north wall of the 'Súra Masjid.' Its lotus ornament shows in design and execution close affinity to the decorative motives of Gandhára sculptures. Though it was evident that this slab had been obtained from some ancient structure in the neighbourhood, my enquiries failed to elicit any indication of its place of origin. The villagers' plea in explanation of their ignorance on this point was that they had come to the place only six years ago when the last redistribution of villages had taken place among the Salárzai clan. The custom here referred to of redistributing at fixed periods the village sites and lands amongst the various sections of a clan by drawing lots prevails, in fact, all through Bunér. It might in itself account to a great extent for the scantiness of local traditions.

Girárai.

There was, however, less difficulty in tracing the ruins about which I had heard at Juvur. They were found to be situated at a place known as *Ali Khánkóte* (' *Ali Khán's* huts '), about 1¼ miles to the west of Girárai. Like the village itself they lie at the foot of the hill range, which divides the valleys of Girárai and Kingargalai. Conspicuous ruins of buildings and terraces, all constructed of ancient masonry, cover the several small spurs which descend here into the valley. The best preserved are on a spur flanking from the west the approach to the gorge through which the direct route to Kingargalai leads.

At the eastern foot of this spur is a narrow tongue of high and fairly level ground, stretching between the bed of the Girárai stream and the entrance of the above-named gorge. On this strip of ground I came upon several circular mounds which are undoubtedly the ruins of Stúpas. The one in the centre still rises to a height of about 20 feet above the ground-level. It has been dug into apparently some time ago by treasure-seekers. The excavation they effected shows the solid, though rough, masonry of which the mound is built. Around it are remains of walls indicating, perhaps, an enclosing quadrangular court. The wall facing west can be traced for a length of 42 feet, that to the north for 40 feet. About 20 yards to the south-west from this Stúpa is another still larger mound thickly overgrown with jungle. It reaches to a height of about 25 feet and has evidently not been disturbed. The remaining portion of the level ground to the east is strewn with small mounds, some of which in all probability mark the site of votive Stúpas of modest dimensions. Regarding a probable identification of this site I must refer to the explanations given below in Section II of this Report.

After returning from Ali Khánkóte and Girárai, I marched along the well-cultivated ground at the northern foot of the hills which separate Girárai and Bampókha. About one mile to the east of Girárai I noticed traces of old walls, much decayed and overgrown by jungle, on a flat terrace-like plot of ground projecting from the hill side. They seemed to belong to a large square enclosure with a stúpa-like mound in the centre. After crossing the broad valley in which the stream coming from the western slopes of Mount Ilm flows down towards Bampókha, I struck the road which leads in the valley of the Barandu River from Bampókha down to Tursák. The dry alluvial plateaus passed along the left bank of the river, the bold and fairly well-wooded ranges to the right towards the Pírsai and Malandri Passes, and the fine view of snowy mountains far off in the Indus direction,—they all reminded me forcibly of scenery I had seen in Kashmír.

Close to the north of the road and at a distance of about 2½ Tursak. miles from Tursak, I found a large square mound rising to about 13 feet above the ground. The late hour of the day at which this interesting site was reached permitted only a rapid examination. It showed that the whole mound was artificial, constructed of rough layers of stone, with masses of débris, apparently from fallen walls, over them. The corners of the mound lie in the direction of the cardinal points. The north-east face, which was more clearly traceable, measured on the top about 100 feet. In the south corner are the remains of a small circular mound which evidently was once a Stúpa. To the south of the latter again and outside the square rises another circular mound about 18 feet high which seems to have been connected with the quadrangular terrace by means of a narrow platform. The position of these mounds is such that the structures marked by them must have been conspicuous objects far up and down the valley when intact. The obligation of arriving in camp before nightfall forced me to leave these interesting remains far too soon. I had hoped that it would become possible to revisit them subsequently from Tursak. In this, however, I was disappointed. It was dark before I reached the camp pitched outside Tursak.

On the following morning (13th January) a column composed of half the Brigade marched from Tursak to the valley of *Pádsháh* in the north. As this move appeared to offer an opportunity for approaching localities on Mount Ilm from which Major Deane's agents had previously procured impressions of inscriptions, I decided

to accompany it.   Before starting I paid a visit to Tursak village
with a view to tracing there the original of the small inscription
which I had published from a cloth impression as No. 27 in my
paper on Major Deane's inscriptions.*   The note which accompanied
this impression described it as taken from " an inscription on a
stone in the wall of the house of a Mulla, Tursak in Bunér. It is
said to have been taken originally from some old ruins with other
stones for building purposes."

On entering the village I soon realized the peculiar difficulties
with which the search for detached inscriptions in Bunér has
proved to be attended.   Neither of the two guides, with whom
Major Deane's kind forethought had provided me, knew anything
as regards this inscription.   I was thus forced to fall back upon
enquiries among the few inhabitants who had not deserted their
homesteads.   None of them could, or would, give information as to
the particular Mulla's house the walls of which must be supposed
to contain this little epigraphical relic.   Tursak is a very large
village, in fact the biggest in Bunér, and boasts among its population of not less than twelve Mullas.   It was with difficulty that
I got half a dozen of these Mullas' houses pointed out to me.   But
the search which I made in succession in these deserted dwellings
proved fruitless, and from the beginning offered little promise.

The walls in the houses examined, like those in most
villages or dwellings in Bunér, are built of rubble and are covered
in large portions with rough plaster.   In several of the houses
there was a number of rooms and sheds ranged behind the
entrance court-yard, indicative of the comparative ease of the
owners.   This meant a considerable addition to the extent of the
wall surface calling for examination.   In order to secure a reason-
able chance of discovering here a small stone, the exposed surface
of which as shown by the impression does not measure more
than 8 by 6 inches, it would have been necessary to scrape the
walls of the plaster wherever it seemed recent, and to devote
altogether to this search far more time than actual conditions
permitted.   The cursory inspection of half a dozen houses and the

---

* See *Journal of the Asiatic Society of Bengal*, 1898, Part I, page 4.

repeated attempts to elicit information from such inhabitants as
the sepoys of my escort managed to get hold of, had already cost
me more than an hour when I turned at last my back on the lonely
alleys of Tursak to start on the march towards *Pádsháh*.

The route leads first to the north through an open fertile
valley, which is watered by the stream coming from Charrai.
Skirting the foot of the high *Jaffar* hill, the road then turns to the
north-east and ascends a low watershed near the village of
Burjo Khána. Here an extensive view opened embracing the
greater part of the fine broad valley of Pádsháh and the whole of
the high mountain range to the north, between the peaks of *Ilm*
and *Dosirri*. The streams which drain this portion of the range on
the south unite close to the village of *Pádsháh*, which thus by its
very position is marked as a place of importance. It is the site of
the holiest Muhammadan shrine in Bunér, the Ziárat of Pír Bába
Sáhib ; it had on this account been singled out for a visit by
General Meiklejohn's column. I had caught up the latter near
Burjo Khána and rode ahead with its advance guard of Guides
Cavalry to close Pádsháh village, which was reached after a march
of about 9 miles from Tursak.

The large Jirgas of the Gadazai tribe, which soon made their
appearance before the Political Officer, showed that, notwithstanding rumours to the contrary, resistance was not to be expected at
this sacred spot either. The troops were accordingly ordered to
halt at *Bhai*, about two miles before Pádsháh, and to return to the
main valley below Tursak on the day following. These dispositions
made it clear to me that my chance of approaching the localities
on Mount Ilm, which had yielded the inscriptions already referred
to, would be limited to the few remaining hours of the day. I was,
therefore, glad to obtain permission to join in the reconnaissance
which Captain Todd, Assistant Field Intelligence Officer, with a
mounted escort was pushing towards the Jowarai Pass to the
north-west of Pádsháh.

At *Lagarpár*, the first village reached, I was able to obtain
accurate information as to the position of *Miángám*, where two of
Major Deane's inscriptions, published by me in the Bengal Asiatic

Pádsháh.

Society's Journal (Part I, 1898, Nos. 29 and 30), had been obtained. It is described as a small village occupied by *Miáns* or Saiyids who have given it its name. It is situated on a shoulder of the great spur which runs down from Ilm Peak in a south-easterly direction. The designation *Ilm-o-Mians* (' Centre of Ilm ') which is used in the notes of Major Deane's agents indifferently with Miángám for the find-spot of these inscriptions, does not seem to be known as a local term, but describes accurately enough the situation of the place. As all my informants agreed in speaking of Miángám as covered with snow at the time, it must evidently lie at a considerable altitude.

Bishunai.

A rough ride of about two miles over a very stony road along the stream which flows from the Jowarai Pass brought us close to the village of *Bishunai*. I had been particularly anxious to reach the latter, as four of the most characteristic inscriptions of the Bunér type, of which impressions have been secured by Major Deane, are described as having been found on stones in the vicinity of this village. They have been published as Nos. 2—5 in M. SENART'S " *Notes d'Épigraphie Indienne,*" Fascic. V.* Having reached so near to the desired point, I felt all the more disappointed when I found that I should have to turn back again without being able to explore it. The escort of Guides Cavalry accompanying Captain Todd was under orders to rejoin their squadron at Bhai in time to allow the latter to return to Tursak the same evening. The time, which remained after the hurried ride up the valley, would barely allow of the ten minutes halt on the road which was required by Captain Todd to sketch the main topographical features of the Pass in front of us. A visit to Bishunai village, which lies a short distance off the road to the north, could under these circumstances not be thought of, still less a search for the inscriptions referred to. For the disappointment thus experienced, the fine view which opened from this point could scarcely afford me compensation. The valley which leads up to the watershed towards Upper Swat, being flanked by snow-covered spurs from Ilm and Dosirri and well-wooded in its higher portion, bore quite an alpine character.

* See " Les récentes découvertes du Major Deane," *Journal asiatique,* 1894.

Returning to Pádsháh as fast as the tired horses could bear us, we passed close to the Ziárat of Pír Bába Sáhib, hidden in a luxuriant grove of Chinars, pines and other trees. An order previously issued prohibited us, like other unbelievers, from entering this the most famous Muhammadan Shrine of Bunér. But the accounts subsequently given to me by those who w‿re allowed to pay their respects to the buried saint, showed that the shrine erected at his resting place can lay claim neither to architectural interest nor antiquity.

The Ziárat occupies a spot close to the confluence of the streams which come from the Jowarai Pass, and the south-western slopes of Dosirri, respectively. The ample water-supply they secure accounts for the evident fertility of the Pádsháh Valley. Both above and below the village stretch broad terraces of well-irrigated rice fields. The well-to-do condition of the place is indicated by the respectable number of Hindu trad rs settled there. Two of these had not fled and were induced to accompany me to the camp at Bhai, where I was able 'to obtain from them curious information regarding the condition of the Bunér Hindus and the sacred sites or Tírthas visited by them in the neighbourhood.

From evidence which I hope to discuss elsewhere, it appears that the Hindu Banias, resident in Swat and Bunér, represent the trading castes of the old Hindu population which had remained in these valleys after the Pathán invasion. Neither they themselves nor their Afghán masters know of any tradition indicating a later immigration from India proper. It is evident that the same reasons which enable these families of Hindu traders at the present day to maintain themselves and their religion amongst the fanatical tribesmen, are sufficient also to account for their original survival. In view of this circumstance it may safely be assumed that the sacred sites to which the pilgrimages of the Bunér Hindus are now directed, mark Tírthas of considerable antiquity.

Tírthas on Ilm.

The most popular of these pilgrimage places seem to be the *Amarakunda* spring and the *Rám Takht*, both situated on Mount

Ilm. The sacred spring appears to lie close to the main summit of the mountain and on its southern face. Remains of an ancient enclosure or building are said to be visible near it. The *Rám Takht* ('Ráma's throne') is described as an ancient walled platform about two miles distant from the Amarakunda and on the northern slope of Mount Ilm towards the Swat Valley. It is visited by the pilgrims in conjunction with the Amarakunda on Sundays falling in the month of Jyaishtha. Sráddha ceremonies are performed at both spots by the accompanying Purohitas, who are said to possess also some account (*máhátmya*) of the legends connected with the Tírthas. Of the few Purohita families of Bunér there are one or two settled at Pádsháh and at Gókand, a village situated some distance further to the north towards Dosirri. But these had fled. I was in consequence unable to ascertain the particular legends which are supposed to account for the sacredness of these spots.

The night from the 13th to the 14th January was passed in bivouac with General Meiklejohn's force in the fields near Bhai village. The troops were to march next morning down to Elai in the Barandu Valley by the direct route leading along the Pádsháh stream. As the information collected by me did not point to remains of interest likely to be found in this direction, I obtained permission to return with a small escort to the Divisional Head-Quarters Camp at Tursak, the neighbourhood of which I had not been able to examine previously. Before, however, starting on the march back to Tursak, I was induced by information given to me regarding certain carved images to ascend the rocky hillside which rises immediately above Bhai to the north-west.

Remains near Bhai. About half a mile from the village and at an elevation of circ. 200 feet above it, I came upon the remains of two Stúpas on a narrow terrace which juts out from the hillside. They are situated close to a spring known by the name of *Jurjurai* and appear now as solid mounds of rough masonry laid in regular courses. The Stúpa immediately to the south of the spring shows a square base, the south-east face of which measures about 50 feet. The height of the whole mound is about 30 feet, but must have been once considerably greater, as the top appears now artificially levelled. About one hundred yards further to the west rises another small Stúpa. Its conical top is comparatively well preserved and shows clearly on its west face the consecutive courses of masonry. The base can no longer be traced distinctly on the hillside. The total height of the mound I estimated at about 35 feet. Traces of old walls and terraces are still visible near these Stúpas.

After climbing some 300 feet higher by a rough path along the steep cliffs I was taken by my Gujar guides from Bhai to a large overhanging mass of rock. This forms on the west a kind of grotto, which seems to have been artificially enlarged. Inside this and on the inner face of the rock, I found a much effaced group of relievos, representing a seated Hindu deity in the middle, with a smaller seated figure on either side. The total breadth of the relievo group is about 5 feet, and the height of the central figure a little over 3 feet. To the right of this group there are two smaller images carved from the rock, each about one foot in height. As all these relievos have suffered considerably owing to the friable nature of the stone, I could not trace with any certainty the deities they are intended to represent. In general style and treatment these relievos seemed to approach closely to the rock sculptures of Charrai described above.

After visiting these remains I marched back by the previous route to *Tursak*, which I reached in the afternoon. Having obtained a mounted escort in General Blood's Camp I then started for a rapid examination of the neighbourhood. The position which Tursak occupies shows great natural advantages. The main valley of Bunér opens there first to greater width and is crossed at this point by a series of convenient routes which connect Upper Swat with much frequented passes leading down to the Rustam Valley. It is evidently due to this favourable position that Tursak is now the largest place in Bunér. The same considerations seemed to indicate that the site was of importance already in earlier times. I was, therefore, not surprised to find that even a

Tursak.

cursory inspection of the neighbourhood acquainted me with ample evidence of ancient occupation.

In the first place my attention was attracted by a series of strongly-built ancient dwelling places visible on the crests and slopes of the rocky spurs of *Jaffar* hill which overlook Tursak on the north-east. They appeared in form and construction to resemble closely the fortified buildings examined near Kingargalai, Juvur, etc. But as they are situated at a considerably greater height above the valley than at the last-named localities, I was unable to spare the time necessary for their examination. Restricting my search to the valley stretching east and south of Tursak, I first visited the village of *Anrapár*, situated on the southern bank of the Barandu River about two miles below Tursak. From there the fertile and well-wooded valley could be overlooked as far down as Dagar.

Stúpa of Gumbatai.

Guided by information obtained at this village, I recrossed then to the left bank of the river and came at the very foot of Jaffar hill, where two projecting spurs form a kind of rock amphitheatre, upon a large ruined site with a Stúpa and remains of a monastery. The former accounts for the name *Gumbatai*, by which the spot is known, *Gumbat* (or *Gumbas*) being the ordinary designation among Afgháns of any ruined building of circular shape, whether a Stúpa, temple or vaulted tomb. The extent of the ruins and their situation only a few hundred yards off the main road, which leads from Tursak to Elai and down the valley, showed clearly the importance of these remains. I accordingly determined after a rapid survey to utilize the following day for their exploration. I returned by nightfall to Tursak, which proved to be only about 1½ miles distant to the north-west by the direct road.

General SIR BINDON BLOOD, to whom I made a report regarding these interesting remains, very kindly agreed to my request and allowed me to employ a small detachment of Sappers on trial excavations at this site. Accordingly on the following morning (January 15th), when the Tursak Camp was broken up and the troops moved off to Dagar and Réga, I proceeded with a small party from the 5th Company, Bengal Sappers and Miners, which the Officer Commanding Royal Engineers could spare from road-making work, to the site of *Gumbatai.*

The ruins as shown in the site plan on Plate No. *III* occupy a broad open glen at the south foot of the Jaffar hill, enclosed in a semi-circle by rocky ridges. The remains now visible above ground form two distinct groups. The larger one lies on a small terrace-like plain at the very entrance of the glen, raised about 50 feet above the level of the river banks. The second group, about 100 feet higher up, is built on the hillside to the north, where the steep slope is broken by a small projecting spur.

At the east end of the lower group rises a ruined Stúpa which in its present state of destruction forms a mound of roughly circular shape, about 55 feet in diameter at its present base and circ. 30 feet high. The level ground immediately adjoining the Stúpa mound in the west is flanked on the north and south sides by two thick walls, 60 feet long, which form a kind of court (marked *A ;* see detailed plan, Plate *IV).* Attached to the west end of each wall is a small circular structure containing a round chamber of 14 feet diameter. Little is left above ground of the walls of these round structures. But from their position and size it can be assumed with great probability that they were intended like the corresponding round chambers in the ruined monasteries of Guniár, Takht-i-Báhi, etc., to serve as chapels for the reception of more important images.

The two walls referred to extend on the east only up to a line which would pass through the centre of the Stúpa. There are no traces of any walls or buildings to the east of the Stúpa, nor of any other structure which could have served to close the Court *A* on this side. The opposite or west side of Court *A* is formed by the enclosing wall of a great quadrangular court (shown as *B* in plan), which almost joins it, the distance between this wall and the circular chapels mentioned being only 15 feet. This court, which is approached by a gate 15 feet broad evidently sighted on the

Stúpa, is remarkable for its size and the massive construction of its walls. It forms nearly a square measuring inside 135 feet in width and 136 feet in length. The walls now traceable above the ground show strangely enough a striking difference in thickness. Whereas they are only 4 feet thick on the north and west side, they measure fully 16 feet in the south and 15 feet in the east. It is probable that this difference must be explained by the thicker walls having been built for the purpose of providing room for small cells, such as are found around the courtyards of several of the Gandhára monasteries and of most of the great Kashmír temples. As the walls inside reach nowhere higher than 4 to 5 feet above the present level of the court, and as the latter has clearly been filled up to a considerable height by the accumulation of débris, the point could be definitely settled only by excavations.

As evidence probably pointing in this direction it may be mentioned that whereas the outside faces of the south and east walls can yet be traced quite clearly rising in many places to 6 or 7 feet above the outside ground level, this is possible only at a few spots in the case of the inside faces. The difference is likely to be due to the greater decay to which the construction of hollow spaces like the supposed cells would have exposed the portions of the walls facing inside. The construction of the walls throughout was found to resemble closely that described above in connection with the Kingargalai ruins. But the size of the stones used was on the whole larger.

In the north-east corner of Court *B* there are walls joining at right angles the north and east enclosing walls. They may have served to form a separate small chapel-court or a dwelling-place. A similar but smaller structure can be traced near the south-west corner of the court.

The second group of ruins higher up the hillside shows in front a walled-up terrace, about 60 feet broad, with a circular structure on one side similar to the 'chapels' flanking the Stúpa Court *A*. Behind the terrace are the remains of walls forming chambers of no great size. About half-way between the two groups of remains I traced an isolated block of masonry about 20 feet square forming a terrace, the original destination of which cannot be surmised with any certainty. A small mound of débris lying near its centre may possibly mark the position of a little votive Stúpa.

After making a general survey of the remains here briefly described, I turned my attention to the Stúpa mound. This, not-withstanding the state of utter dilapidation to which it has been reduced, still reaches to a height of about 30 feet above the present ground-level. The mass of rough masonry of which the Stúpa was constructed has evidently been used for a long time back as a conveni[.]nt quarry. On the north face regular courses of large blocks could still be clearly distinguished ; the other sides of the mounds are hidden by large masses of débris. No clear idea could thus be formed of the original shape of the upper portion of the Stúpa.

Excavation at Gumbatai.

The centre of the mound has been dug into from above to a depth of about 10 feet. Judging from the comparatively thin growth of jungle on the south face where most of the materials then extracted had been thrown down, the digging could not have been done many years ago. The treasure-seekers, who were then at work here, had evidently not carried their labors deep enough to touch the main deposit of relics which from the analogy of other Stúpas may be supposed to be placed on or below the level of the base.

In order to obtain some indications as to the position of the Stúpa base and the depth of the original ground level in the court, I had trial trenches opened by the small party of Sappers, both at the west entrance of Court *A* and at the foot of the Stúpa mound to the west. At the latter place the Sappers after working through about 3 feet of débris came upon a solid block of closely grained stucco which when cleared was seen to mark the corner of a square platform. The exact spot at which this corner was struck is marked by *c* on the plan. The block

forms a square of 9 inches with a height of 13 inches. It is
ornamented on two sides which were found to face nearly due
west and south. That this was the original position of the block
was made evident by a stone base unearthed below it which
showed exactly the same bearings.

The little stucco pilaster is ornamented at its foot by a series
of mouldings. These project about 1 inch beyond the flat middle
portion of the block which is about 4 inches high. The top part,
about 5 inches high, also projects and shows a kind of egg and
dart ornament in bold relievo and in two rows divided by a narrow
band. The stone base below the stucco-block could be cleared
only to a depth of about 10 inches. Its top forms a square of 1½
feet, and is decorated on the sides facing west and south by a bold
cornice projecting in several well-carved mouldings to a total
breadth of about 5 inches. Continuing the excavation to the east
of this corner and towards the Stúpa for a distance of about 5 feet
a masonry wall was laid bare running flush with the south face of
the stucco-pilaster and its base. Fragments of stucco were found
sticking to the joints of the masonry courses. It may thus
be concluded that this wall was decorated similarly to the above
described corner.

From the position occupied by this wall, as shown on the plan,
it will be clear that it could not have formed part of a square
basement of the whole Stúpa. It is more likely to have belonged
to some platform raised by the side of the Stúpa and possibly on
the basement of the latter. Such a platform might by the analogy
of the examples presented in the ruins of Takht-i-Báhi, Jamálgarhi
and other Gandhára monasteries (see *Arch. Survey Reports*,
V., pl. vii, xiv) be conjectured to have served either for the
placing of images or a small votive Stúpa. In support of this
conclusion reference may also be made to the comparatively high
level at which this stuccoed wall was unearthed. Near the west
entrance of the court the present ground level seemed lower
than at the foot of the mound. Yet a trial trench carried

down to a depth of fully five feet, failed to reach there the original floor of the court. The accumulation of débris must be supposed to have been even greater immediately around the Stúpa. There is thus reason to assume that the real base of the Stúpa is yet buried at some depth below the platform brought to light. This will also explain why the ornamented stucco-pilaster remained the only piece of sculptured work unearthed during this brief excavation.

I regret all the more the very limited extent of the excavations made, as the explanation given below, page 61, will show that these ruins may be identified with great probability with a sacred site of considerable fame described by the Chinese pilgrims. The Sapper detachment had orders to follow at no great distance the rear guard of the column which was moving down the valley to Dagar. The men were accordingly obliged to stop digging early in the afternoon. I myself left Gumbatai some hours later after completing the survey of the ruins, in order to rejoin General Meiklejohn's Camp at Réga. I first marched in the fertile plain by the left bank of the Barandu River to a point about two miles lower down the valley. From there the village of *Elai*, picturesquely situated in the angle formed by the Barandu and Pádsháh rivers, could clearly be seen. From the hillside above Elai some small inscribed stones, now in the Lahore Museum, have been picked up by Major Deane's agents. According to the information supplied to him there were no ruins near. The distant view of the hill slopes with which I had to content myself, also failed to show me any ruined buildings above ground.

Elai.

I then crossed to the right river bank and followed a track leading over an arid alluvial plateau to a point about two miles distant from Elai where the river passes through a remarkably narrow gorge of sandstone rocks known as the *Survai Khandau*. A short distance below this gorge the road to Karapa and the south-eastern portion of the valley known as Panjpao turns off to the right. It ascends a narrow and steep defile which cuts through

the rocky range of high hills lining here the south or right bank
of the Barandu River. Through the whole of the gorge, which is
about three-fourths of a mile long, there leads a fairly broad path
fit for laden animals. It is cut either into the rock or carried on
walled-up foundations of ancient masonry along the cliffs.

Karapa Road.

This road, which in its construction, resembles closely the
ancient roads over the Malakand and Sháhkót Passes already
mentioned, goes back undoubtedly to pre-Muhammadan times. It
may safely be taken as an indication of an important route having
led already at that period through the Karapa defile. The latter
is crossed by the direct lines of communication connecting the
Malandri and Ambéla Passes with the central portion of Bunér
and hence with Pádsháh and the other routes into Upper Swat.
The above-named passes must have at all times attracted traffic.
They give access to the old trade emporium marked by the site of
the present *Rustam*,* and to the important ancient route leading
to the east of the Indus viâ *Udabhánda* (Waihand, Und) and
Taxila. The evident care bestowed on the construction of a road
through the naturally difficult Karapa defile which falls into the
direct line continuing those routes to the north, is thus easily
accounted for.

After crossing this defile the large village of Karapa was passed
at the edge of the Panjpao plain. Proceeding about two miles
further to the south-east I reached after nightfall the village of *Réga*,
nestling at the entrance of a side-valley, and the camp established
there The village had been singled out for a visit of General
Meiklejohn's Brigade as the home of the "Mad Fakír" whose
fanatical preaching had been the immediate cause of last summer's
rising in Lower Swat, the siege of Malakand, and the events that
followed. After assisting in the early morning of the 16th
January at the destruction of the Fakír's house and mosque which
were blown up and burned, I proceeded to the examination of
the ancient remains reported to me in the vicinity of *Sunigrám*.

*See Gen. CUNNINGHAM's Ancient Geography, pp. 65 sq.

Major DEANE had already previously heard of them through one of his agents As this agent (Sherbáz) now actually accompanied me, I had no difficulty in finding the ruins referred to.

About one mile to the north of Réga where the valley leading down from the Malandri Pass in the south-west debouches into the Panjpáo plain, I had already on the previous evening when on my way from Karapa to Réga, noticed a large mound suggestive of the remains of a Stúpa. This assumption soon proved correct on closer inspection The mound rises to a height of about 25 feet above the flat level of the plain. Wherever the débris covering its sides had been removed by the action of rain or other causes, it showed the same courses of rough masonry which had been noticed in the Stúpas previously described. Judging from the dimensions of the present base of the mound which measures circ. 240 feet from east to west and 200 feet from north to south, this Stúpa must have been by far the largest of all those examined in Búner. If a conclusion can be drawn from the state of utter dilapidation in which it is now, it may also be looked upon as one of the oldest. At about half its height a kind of terrace can be traced all around the mound : this probably indicates the elevation from which the Stúpa proper rose above the basement.

The top of the mound now forms a slightly sloped oval measuring circ. 120 feet from east to west and 75 feet from north to south. I am inclined to explain this peculiar shape by the assumption that the basement which shows a similarly elongated form was broader to the east and west than on the other two sides. The decay of the originally hemispherical mound must thus have been more rapid on the north and south sides where there was no broad terrace to retain the loose masonry brought down by the rains, etc., than on the east and west where the masses of débris accumulated over the original basement. In support of this explanation I may mention that the slopes of the mound to the north and south appeared steeper. It is just on these sides that the courses of masonry composing the mound are traced most clearly on the surface.

Stúpa of Sunig.Am.

For some distance from the foot of the mound to the south the ground is covered with low heaps of débris which seem to indicate the site of ruined buildings once attached to the Stúpa. These remains were, however, too indistinct and too much overgrown by jungle to permit of a plan being taken in the short time available.

Well near Sunígrám. At a distance of about 60 yards to the south-east of the Stúpa there is an ancient stone-lined well which has remained on the whole in a remarkably good state of preservation. The well proper is 8 feet in diameter and is enclosed by a circular wall, 5 feet thick, of carefully set masonry. Adjoining to the west is a staircase which leads between equally well-built walls down to the level of the water. This is now 18 feet below the ground level, and is reached by 23 steps. The accompanying plan and section (V) shows the construction of the well. Some of the stairs have crumbled away, and also the side-walls have suffered in parts notwithstanding the repairs which are indicated in several places by coarse masonry of a later date.

Apart from these repairs the whole of the walls shows to perfection that peculiar form of masonry—large blocks in level courses and columns of small stones in the interstices—which has been described already above as characteristic of all the ancient structures in this and the neighbouring regions There is no special feature to indicate the relative age of the well as compared with that of the ruined Stúpa. Its escape from the fate of the latter may be due to continued use and consequent repair. Some Khattrís from Réga whom I met near by were prepared to ascribe the well to *Birmal, i. e.,* Birbal, the renowned minister of Akbar. But this tradition, if it is one at all, cannot refer to anything more than a clearing of the ancient well which may have become disused and filled up with earth. These informants knew of no other name for the site but *bahai,* which in Pashtu is the ordinary designation for any stone-lined tank or well with steps leading down to the water.

Sunígrám. The village of *Sunigrám,* a small place, lies about half a mile to the north of this site. It occupies a saddle-like depression between the east foot of the rocky hill range through which the Karapa defile leads, and a series of small fir covered hillocks which rise like islands from the plain and form a continuation of that range to the south-east. There is nothing ancient to be noticed about the village but its name *Sunigrám*, which is undoubtedly of Indian origin and hence old.

The second part *grám*, from Sanskrit *gráma* ('village'), does not occur in any other Bunér local name I know, and is but rarely met with in the neighbouring territories of Swat and Yusafzai (see *Jolagrám*, *Pajigrám* and *Udegrám* in the Swat Valley, *Naugrám*, on the Khudu Khel border, *Asgrám* and *Kábulgrám* on the Indus). It is scarcely necessary to point out how common on the other hand this ending, in its varying vernacular forms of *grám*, *gám*, *gáon*, *gráon*, etc., is throughout the whole of Aryan India. The first part of the name *Suni*—is clearly connected with Sanskrit *suvarna*, 'gold,' and represents probably a Prakrit derivative of *sauvarnika*, 'goldsmith.' Thus in Kashmírí, which may be considered a near relative of the old Indo-Aryan dialect once spoken beyond the Indus, we have *sun* ('gold') and *sunar* ('goldsmith') derived by a regular process of phonetic conversion from Sanskrit *suvarna* and *suvarnakára*, respectively (compare also Hindi *suniyár*). Derivatives from Sanskrit *suvarna* are not amongst the words borrowed by Pushtu from Indo-Aryan dialects. It is thus certain that the local name *Sunigrám* must go back to a period preceding the Pathán occupation.

Immediately above the village, and to the west of it, rises the rocky hill range which has been mentioned in connection with the Karapa defile. Guided by Sherbáz, I ascended its steep scarp in a northerly direction to a height of about 400 feet above the bottom of the valley until I reached the point from where a rocky spur running south-east to north-west juts out towards the Barandu River. It is about one-third of a mile long and is known by the name of *Panjkótai*. The crest of this spur is fairly level and bears the ruins of a large number of buildings which in construction and character resemble closely the ancient dwelling places examined near Kingargalai and Juvur.

Panjkótai.

On the west slope of the spur and towards its north-west
extremity overlooking the river are the comparatively well
preserved ruins of what evidently was once a monastery of great size
and importance. They consist, as shown on the attached site-plan
VI, in the first place of a series of large terraces. These are built
against the hillside by means of strong supporting walls and extend
for nearly 300 feet from north to south with a total breadth of over
160 feet. At the south end of these terraces rises a block of vaulted
rooms with attached courts constructed of solid and carefully set
masonry. At the north end of the terraces and close to their edge
are the much injured remains (*B*) of some smaller structures.
Among them is a square block of masonry (*C*), which judging from
the remains of a small circular mound built over it can have been
nothing but the base of a little Stúpa. The circular pit excavated in
the centre of this mound shows that treasure-seekers have ere now recognized its true character and been at work here. The little square
structure (*D*) to the east, which is even more injured, may also
mark the position of a small Stúpa.

Vihára of
Paajkótaí.

The interest of the main building *A* lies in the good preservation of its superstructures which acquaint us with some details of
architectural construction not otherwise traceable in the extant
remains of Bunér. They are illustrated by the detail plan *VII*.

Three rooms of this building forming its south and west side
show high pointed vaults of overlapping stones which spring
from a projecting cornice of the longer side walls. The height from
this cornice to the point of the arch is 10 feet 3 inches. The construction of the vault and cornice is shown by the section given for
the line *cd* in the detail plan   The total height of the rooms could not
be ascertained, as the interior is partly filled up by masses of débris
from the fallen portions of the vaults. The width of the vaults is
11 feet in the two larger rooms *E* and *F* which are 31 and 35 feet
long, respectively. In the small room *G* the width covered by the
vault is 8 feet 3 inches and the length 12 feet. These three rooms
communicated with each other and the central court *H* by means of
passages of varying width surmounted by pointed arches of

overlapping stones. Owing to the accumulation of débris only these arches are now visible above the ground level. The elevation of two of these passages, *X* and *Z*, has been shown in the detail plan.

Besides the passages leading into the central court there were windows to admit light into the larger rooms. These windows are placed in the centre of the walls and end in pointed arches, as shown in the section of line *cd*. Their width is 2 feet, and their height to the point of the arch is 5 feet 6 inches. The sill or lower edge and the sides of the windows are bevelled inwards by means of regularly receding courses, evidently with a view to distributing the light more evenly over the room.

The central apartment *H* is the largest in this pile of building, being 35 feet by 16 feet. It does not appear to have been roofed, and must hence be assumed to have formed a kind of central courtyard. It has no direct entrance from outside, but was evidently approached by a passage which leads to the adjoining room *I* through a dividing wall now for the greater part broken. This latter apartment, which is also 35 feet long, but only 9 feet 3 inches broad, does not show either any trace of having been roofed. In its southeast corner are the remains of a staircase *e* leading up to the open quadrangle *J* which occupies the raised terrace immediately to the east. As this staircase, as far as can be judged from the present condition of the building, was its only entrance from outside, we may conclude that the small court *I* formed a kind of open anteroom to the whole block.

The quadrangle *J* is in reality a terrace, 55 feet deep and 50 feet broad, built against the rising slope of the ridge and screened on the east and south by strong walls 7 feet thick. The unusual thickness of these walls suggests that they contained niches which might have been used as small cells. But the ruined condition of the walls and the great masses of débris and earth which cover their foot inside the quadrangle made it impossible to ascertain this point. Judging from the relative position and size of this enclosure, it might be conjectured that it served, like similar open courts in the ruined monasteries of Takht-i-Báhi and Jamálgarhi, described by Sir A. CUNNINGHAM (*Archæological Survey Reports, V*, pages 30, 50) as a meeting place for the fraternity of monks.

To the north of this court, but at a considerably lower level, extends another large platform (*K*), 110 feet long, which shows no trace of superstructures. From 'this a flight of 7 steps leads down to the artificially levelled ground on which the main block of building stands. Immediately to the north of the latter is a large terrace, 103 feet long by 88 feet broad, supported on the sides facing the downward slopes of the hill by basement walls over 30 feet high.

Construction of Panjkótai ruins.

The massive construction of these walls and the great exent of the terraces which they support suffice to indicate the importance of the site. The blocks of stone used in the walls, both of these terraces and of the main building, are in general larger than in any other structure examined in Bunér; they are often over 4 feet long with a thickness of 1 foot. Though the blocks are on the whole but roughly hewn, as throughout the masonry of the ancient buildings in Gandhára and Udyána, yet special care has been taken to arrange them in even and regular courses The interstices of each course are not merely filled as usual with closely packed columns of small flat pieces, but show besides the use of a kind of thin mortar which must have added considerably to the consistency and strength of these walls. It is evidently due to the exceptional solidity of the construction that the walls of the main block still show a height of 23 feet at the north-west corner where they rise on the massive foundation of the terrace basement.

An equally significant feature of the *Panjkótai* ruins is the comparatively great span of the overlapping domes which form the roofs of the two large rooms in the main building. The span of 12 feet covered by these domes is not reached by any extant arched structure in Gandhára or Udyána. The domes in nearly all the buildings surveyed by General Cunningham are limited to about 8 feet (see *Archæological Survey Reports*, *V*, page 52). The wider span assumed by him in two examples is a matter of conjecture.

It will help us to form a correct estimate of the relative importance of the Panjkótai ruins, if we compare them also in other respects with the remains of such well-known sites as Takht-i-Báhi and Jamálgarhi. This comparison suggests itself all the more, as the general situation of the ruins near the ridge of a steep rocky spur bears a striking similarity to that of the last-named great monastery. Referring then to the plans of the latter, as recorded after excavation in plates vii. and xiv. of General CUNNINGHAM'S *Archæological Survey Reports*, Volume V, we note at once that though the number of separate buildings at present traceable at Panjkótai is far smaller than that brought to light in the course of prolonged explorations at the above two sites, yet the size of the structures still above ground at Panjkótai is decidedly more imposing.

The same holds good as regards the extent of the terraces and their substructures which here as there were indispensable to provide the requisite level building ground. That the ample space thus provided at Panjkótai was once occupied by a greater number of buildings than now visible can be inferred from the low mounds of débris which stretch in various directions across the terraces to the north of the main pile of building. It is likely that these little mounds, of which, I regret, it was impossible to make any plan in the very limited time available, mark the position of small detached structures which here as at Jamál-garhi may have contained the cells of the monks attached to the establishment. Other small buildings of this kind situated nearer to the rising slope of the ridge are, perhaps, buried under the masses of detritus carried down from the latter.

It remains yet to be noted that the Stúpa *B* referred to above would, as shown by the dimensions of its extant base, 25 feet square, well bear comparison with the corresponding structures of Takht-i Báhi and Jamálgarhi. The chief Stúpa of the first-named monastery rose on a basement, 20½ feet square (*Archæological Survey Reports*, V., p. 26), and the 'great Stúpa' of the second did also not measure more than 22 feet in diameter (*ib.*, p. 47). The oblong enclosure *D*, 30 feet long by 20 feet broad, which adjoins the Stúpa of Panjkótai on the north may like the small 'Chapel courts' found at the two Gandhára monasteries, have served for the placing of Buddha statues. But the walls of this enclosure are in so ruined a

condition and its interior so much covered with débris that any
conjecture regarding its original character, if not tested by excavation, must necessarily remain hazardous.

I cannot conclude this account of the ruins examined on the
Panjkótai spur without referring to the magnificent view enjoyed
from their site. Standing at the north-west corner of the walled-up
terraces, near the remains of the Stúpa, I had before me the whole
expanse of the Barandu Valley stretching with a varying breadth
of 4 to 6 miles from *Elai* in the west towards *Matwanai* in the east.
The river, which winds along the southern side of the valley often
divided into several channels, passes close to the north foot of the
spur. Looking to the north beyond the valley and the hill range
immediately skirting it, the double-peaked cone of Mount *Dosirri*
with its cap of snow came prominently into view. To the north-west
the fir-clad slopes of Mount *Ilm* could clearly be seen through the gap
formed by the valley which runs down to Elai. In the west
appeared the rugged heights of *Jaffar* hill near Tursak. In the east
the extensive view across the plains and low alluvial plateaus of
the central Barandu Valley was limited only by the high
Dúma range which divides Bunér from the Indus Valley. From
the steep cliffs, which form the extremity of the ridge towards
the river and overlook the ruins, the panoramic view was still
wider. It comprised the long—stretched ranges which run up
towards Mount Mahában in the south-east, and the still higher peaks
of the 'Black Mountains' beyond the Indus.

The prominent position occupied by the Panjkótai ruins and
their relatively great extent are indications that the convent to which they belonged must have been once important and well known. It is necessary to lay stress on the
evidence furnished by these points. It will help to strengthen
the arguments set forth below regarding the probable identity of
these remains with the *Mahávana* monastery of Hiuen Tsiang (see
below, Part II, p. 60).

Takhtaband
Stúpa.

Already when standing on the height of the spur above the
Panjkótai ruins a massive mound of masonry further down the

valley had attracted my attention. It was the Stúpa which Sherbáz's report had led me to expect in that direction. To this I
proceeded accordingly when the survey of the Vihára remains
was completed. From the foot of the spur the way lay across the
level plain which stretches here on both sides of the river. After
going for about 1½ miles in the direction of E. S. E. and crossing
the river I reached the Stúpa.

It rises a short distance from the left river bank, about 60 yards
from the extreme western end of a small rocky ridge which without
attaining any great height stretches across the valley to the east towards Shalbandai. This Stúpa, which from the name of the
village nearest to it on the right bank of the river I propose
to call that of *Takhtaband*, has even in its present damaged
condition better retained its original appearance than any other
structure of this type in Bunér. It forms a dome of a shape
somewhat resembling that of a bulb and rises to a height of
about 26 feet above its base. It is constructed of horizontal courses
of massive but rough masonry, none of the stones now exposed
having received any dressing.

This dome is again raised on a large base about 25 feet high,
which originally formed a square measuring about 84 feet at the foot,
approximately orientated. The accompanying elevation (Plate *VIII*)
shows the exact dimensions of the Stúpa and its basement. Both
have completely been stripped, evidently long ago, of their outer
casing of masonry. No remains of it can be traced now on or about
the mound. It was evidently carried away to be used as building
material. In the same way the inner masonry has also been cut
away to some depth round the foot of the Stúpa, the upper
portions of which in consequence are now overhanging.

The Stúpa has been opened by a broad cutting which reaches
to the centre and runs through its whole height on the east side.
This excavation has been carried even further down into the base
to a depth of about 8 feet. There can thus be no doubt that relic
deposits have been reached and abstracted. It must be supposed
that this spoliation took place a considerable time ago as the
débris of the materials excavated can no longer be distinguished.

The cutting here indicated has laid bare a little chamber lined with large and carefully cut slabs in the centre of the Stúpa. It is 7 feet high and forms a square of 7 feet, of which the eastern side is now removed. The floor of this chamber was originally about 12 feet above the level of the Stúpa base. There is every reason to believe that this receptacle was intended for a relic deposit. Square hollows or wells of exactly similar position have been found in several of the Stúpas excavated in the Punjab and the Kabul Valley, also in the great Stúpa of Manikyála.* As far as I could examine the walls of this chamber from below they bear no trace of any decoration or inscription. In order to reach them closely a ladder or scaffolding would have been necessary.

The elevation reproduced shows that there must have been a platform extending round the foot of the Stúpa which had served as a procession-path. But owing to the dilapidated condition of the base, the original width of this platform can no longer be ascertained. It is probable that it was approached from the east, as on this side there are traces of projecting masonry which may have served as the substructure of a staircase.

Neither in the narrow flat gap, which separates the Stúpa mound on the east from the foot of the rocky ridge above referred to, nor on the open ground on any of the other sides was I able to discover any remains above ground which might indicate the previous existence of walls or buildings. It must, however, be noted that the ground all around the Stúpa, which is of a rich alluvial soil, is under cultivation. This would easily account for the removal of such remains if they were not of a very massive character. The late hour at which I reached this site and the necessity of returning soon to camp did not allow me to examine the slopes of the ridge closely. It is possible that remains of dwelling places for the attendant priests could be traced there. From below none were discernable.

---

* Compare Gen. CUNNINGHAM, Archæological Survey Reports, V, pl. xxii.

During the day a portion of the Brigade had marched at no great distance down the valley to Bájkatta. To this circumstance was probably due the utterly deserted condition of Takhtaband village, where I was hence unable to obtain any local information regarding the Stúpa.

The night was passed in General Meiklejohn's Camp near Barkili, which I reached after a march of about 3 miles from Takhtaband. There I ascertained that the greater portion of the force was to move on the following day into the Chamla Valley *en route* for the Ambéla Pass. This was probably the last day I could hope to spend on the soil of Bunér proper. I accordingly resolved to utilize it for an attempt to reach the sites near the villages of *Nawakili*, *Mullaisap* and *Zangi Khán Banda* from which a number of inscriptions either in original or impressions had been obtained by Major Deane's agents. These villages, all belonging to the Nórizai clan, are situated in the valley which leads from Karapa in a south-westerly direction to the Malandri ' Pass.

Starting in the morning of the 17th January I marched first round the foot of the several spurs which descend from the high range to the south and run out into the Panjpao plain between Barkili and Karapa. On the way from the former place to Réga I passed the opening of the valley known as *Béshpúra*, evidently an old name of Hindu origin to which Captain F. S. Robertson, of the Survey Department, had been kind enough to draw my attention. The valley is now practically uninhabited. At Karapa, which is a thriving village of some size, I picked up Aslam Khán, one of the inhabitants, who had assisted my guide Katór Sháh on previous occasions in tracing inscribed stones in this neighbourhood. He first offered to show me "Búts" on the hillside west of the village. But after reaching the small cave to which I was taken, and examining with some difficulty its narrow recesses, I convinced myself that the supposed relievo images were only natural markings of the rocks.

Barkili.

I then marched in the broad open valley to the south-west until at a distance of about 4 miles I reached *Nawakili*, a fair sized village situated at the point where the valley forms an inlet to the south towards Mount Guru. About half a mile to the south of the village is a mound covered with old masonry known as *Surkhau Kandar*. It occupies the west foot of a small fir-covered spur, and on the sides seems partly to have been terraced. On the top old walls are clearly marked. The centre is occupied by a square of old masonry, 34 feet each face, rising only one or two feet above the ground. The western face is continued to the south by another wall for about 22 feet, and this is approached by a kind of terrace sloped as for stairs.

It was here according to Katór Sháh's statement that he picked up, from below the north face of the mound, one of the inscribed stones delivered to Major Deane. Of another stone said to have been found further down the slopes, the agent who accompanied Katór Sháh on that occasion is supposed to have taken an impression.

I was particularly anxious to ascertain the position of the large inscription in unknown characters, of which an impression, marked as having been obtained at Nawakili, had reached me from Major Deane in September 1896. It is now reproduced on No. 82 of Plate X in my second paper on these inscriptions. But the villagers whom I examined would know nothing either of this or any other inscribed stone in the neighbourhood. Aslam Khán who, I have reason to believe, acted as guide to at least one of Major Deane's agents in this vicinity, grew equally ignorant in view of this attitude. After repeated attempts to elicit information by various means I was reluctantly obliged to abandon the search.

The motives of the villagers in denying all knowledge of inscriptions are not far to seek. Their combined fanaticism and ignorance must make them anxious to keep from the 'unbeliever,' in particular when he appears as one of the invaders, information about records which might be supposed to lead to the discovery of hidden treasure or similar advantages. Obstacles of this kind could, among a population as fanatical as the Bunérwáls, be overcome only by the fear of a more immediate danger. But in the present circumstances, when the evacuation of the territory by the troops was known to be a matter of a few days only, the threat of more stringent measures, even if I had been able to give effect to it, would have probably produced no result. It was but too clear that, with an escort of eight sepoys and the certainty of the near retirement of the troops, little impression could be made.

The advanced hour and the necessity of reaching before nightfall the distant camp at Barkili obliged me to forego a visit to *Zangi Khán Banda*. This place from which a series of stones inscribed with very peculiar characters had been secured at several occasions by Major Deane's people,* was according to local information at a considerably greater distance towards the Malar:Jri Pass than the available sketch maps had led me to suppose. Nor could I have reasonably expected to fare there better than at Nawakili, seeing that even Katór Sháh denied having had anything to do with the finds in that locality.

Zangi Khán Banda.

Marching then back from Nawakili I took occasion to visit *Mullaisap* (for Mulla Isuf ?) which lies in a side valley opening to the south-east, about half way between Nawakili and Karapa. Two impressions had reached me of inscriptions near this village. But my local enquiries as to the actual position of the stones were here also of no avail. I could, however, convince myself that neither here nor at Nawakili nor at Karapa were there any conspicuous ruins with which these inscriptions could be connected. On the other hand, none of the sites, at which remains of Stúpas or monasteries are still extant, have hitherto contributed to our

Mullaisap.

* See Nos. 47—50 of the inscriptions reproduced in Part I of my " *Notes on new inscriptions discovered by Major Deane.*" According to the information supplied with them these stones were "dug up from what appears to be an old Memorial Stúpa completely buried in the ground at *Bughdarra*, which is the ravine near Zangi Khán Banda. " For other inscriptions from this locality see Nos. 79·81 of Part II.

collection of Bunér inscriptions. This observation seems to give some foundation to the belief that the originators of the latter must be looked for elsewhere than among the founders or attendants of the Buddhist shrines still extant in ruins.

I reached Barkili Camp, where only a small detachment of troops had been left, late in the evening, having marched my escort that day probably not less than 25 miles. On the next day, the 18th January, the remainder of the troops still in Bunér was under orders to retire over the so-called Bunér Pass and to join the 2nd Brigade which had in the meantime occupied the head of the Chamla Valley through the defile of Ambéla. In order to utilize the few hours still available to me on Bunér soil I moved in the morning in a northeasterly direction down to the river. There an isolated hill rising several hundred feet from the plain close to the villages of Kalpanai and Bájkatta offered a central and very comprehensive view over the whole of Lower Bunér. From Matwanai in the east, where the Barandu River enters a narrow defile leading down to the Indus, to Elai in the west the whole expanse of the valley on both sides of the winding river lay clearly before me. No ruins or artificial mounds offered themselves to view from this commanding position, except the Stúpa of Takhtaband already described. Nor could the Hindu traders, whom I got hold of in Kalpanai village, tell me of any other ancient sites within reach besides those already visited.

Chamla
Valley.

I accordingly returned by midday to the deserted camp of Barkili and hence crossed with the rear guard the pass usually designated as that of Bunér which leads to the head of the Chamla Valley. The latter is drained by the river, which receives the streams from the northern slopes of Mount Mahában and joins the Barandu not far from its own junction with the Indus. Chamla geographically as well as ethnographically forms a territory distinct from Bunér proper. The fir-covered top of the pass was reached through very pretty forest scenery, and offered to me once more a striking view across Bunér, bounded in the north only by the snow-capped ranges of the Dúma Mountains, Dosirri and Ilm.

Reaching in the afternoon the camp which was pitched below the village of Ambéla, I took an opportunity to represent to General JEFFREYS, Commanding the 2nd Brigade, my desire of approaching Mount Mahában as closely as the military dispositions permitted. From the time that the Bunér Expedition had been taken into view I had fondly entertained the hope that it would give me the chance of reaching that mountain which has never yet been visited by a European or surveyed. This desire arose from the fact that of the various positions which have been proposed for the *Aornos* of the historians of Alexander there is none which in my opinion has a better claim for serious consideration than Mount Mahában.

I need not review here the numerous opinions which have been advanced since General Court took up the question in 1836 regarding the site of that famous mountain stronghold. They have been last fully set forth and discussed by General CUNNINGHAM in a separate chapter of his "Ancient Geography of India."* Nor is this the place to explain the reasons which seem to me to militate against any one of the suggested sites that are at present accessible for examination, such as ' Rája Hodi's Castle' opposite Attock, the Karamár hill, the ruined castle of Ránigat. †

The claims of Mount Mahában were first advanced by ˙the late General ABBOTT, of Abbottabad, nearly half a century ago They were rightly based by him on the close agreement which the main orographical features of that mountain, as then known, its proximity to the Indus, its great height and extent, present with the description of

Mount Mahá-ban.

* See pages 58 *sqq.*; compare also Sir E. Bunbury's *History of Ancient Geography*, I, pages 496 *sqq.*

† General Cunningham himself evidently after a good deal of hesitation settled upon Ránigat as the most likely position. But that distinguished antiquarian, to whose intuitive perception in matters of ancient topography we owe many happy identifications, was himself constrained to own in this case that he did not feel satisfied with this location. To any unbiassed student of the question who has visited the ruins on the Ránigat hill, the objections must appear unsurmountable. Its great distance from the Indus, its comparatively small height and still smaller summit are all features which cannot be reconciled with the salient points of the Greek accounts.

the Greek historians.* No fact has since come to light which
could shake the weight of the arguments derived from this
observation. † But the heights of Mahában have continued
to be as inaccessible to Europeans as they were then. It has hence
been impossible to obtain that detailed topographical evidence,
without which it seems hopeless to expect a definite settlement
of this much vexed question.

My interest in Mount Mahában as the probable site of Aornos
was considerably increased by the important information which
Major DEANE had recently obtained through native sources regarding
extensive remains of an ancient fort situated at a point of Mahában
known as *Sháhkót*.‡ Can these ruins be referred to so early a
date as Alexander's invasion, or do they at least indicate the
likely position of an older fortification? Only an archæological
survey of the mountain could give us the answer.

The ready submission of the Chamla tribes induced
the military authorities to abstain from any further advance to the
east down the Chamla Valley. This made it clear to
me that the hope I had cherished of visiting Mount Mahában could
not be realized on the present occasion. If the head of the Chamla

---

* See General Abbott's paper "*Gradus ad Aornum*," *J. A. S. B.*, 1854,
pages 309 *sqq*. Before him General Court already seems to have thought of
Mahában as a possible position for Aornos ; see his incidental reference, *J.A.S.B.*,
1839, page 310.

† The main-objection which General Cunningham raises to Mount Mahában
as the representative of Aornos (*Ancient Geography*, page 61 *sq*.) is based on
the assumption that it is the 'great mountain' by the side of which the *Mahávana* monastery of Hiuen Tsiang was situated. "If any fort had then existed
on the top of the mountain," General Cunningham argues, "it is almost
certain that the pilgrim would have mentioned its name," etc. After what we
have shown below as to the real position of the Mahávana convent, it is clear
that this negative argument, weak in itself, falls to the ground.

Nor can I attach any greater importance to his other two objections, derived
as they are from such defective information as has hitherto been available regarding the shape and extent of the mountain and its several spurs. In the
absence of any proper survey it is impossible to assert the easy accessibility of
the mountain as contrasted with the description given of the steepness of Aornos,
or to compare its circuit with the varying figures recorded for the latter by
the historians of Alexander.

‡ See his above quoted paper, *J. R. A. S.*, 1896, page 673.

Valley had been occupied for more than a few days, the despatch
of a separate detachment to that distance. might yet have possibly
been arranged for in the interest of the topographical survey. For
this Mount Mahában owing to its height and position represents
also a point of considerable importance. But the evacuation of
Chamla and the return of the whole of the force to British territory
were already fixed for the following day. The hopes of Captain
Robertson, the Field Survey Officer, were like my own doomed to
disappointment.

Considering the circumstances I could but feel grateful when
General JEFFREYS very kindly agreed to let me utilize that last day for
a rapid excursion down the Chamla Valley. In order to enable me
to extend it as far as possible he was pleased to grant me a mounted
escort from the Xth Regiment Bengal Lancers. I had thus at least
the satisfaction of approaching the north foot of Mahában closer
than I could have hoped otherwise.

Starting on the morning of the 19th January from the camp   Súra.
below Ambéla I reached after a ride of about four miles the large
village of *Súra* situated on the southern side of the valley. Some
Hindu Khattrís of this place whom I examined knew of an old site
about half a mile to the south of the village and at the foot of a low
spur which descends here from the Sarpati Range. On proceeding
to it I found a spring enclosed in a square basin of ancient masonry.
This is visited as a Tírtha by the Hindús of the neighbourhood.
Close to the west of the spring is a terrace-like mound about 20 feet
high, the upper part of which appeared artificial. The top, which
formed a small plateau about 200 feet from west to east and 100 feet
broad, is covered with remnants of old walls built of large but
undressed stones. There are evident traces of a terrace about
15 feet broad which seems to have run round the mound at a lower
level. The sides are covered with broken pottery. I was unable
to ascertain from my Hindu guides any tradition regarding this
site, or the special name of the locality. They too were well
acquainted with the sacred Tírthas on Mount Ilm and had more
than once performed the pilgrimage.

I rode on through the level ground of the valley, which is here
more than a mile broad. and well cultivated, past Nawagai and
Timúli Dhérei, until I reached the small village of *Katakót*. There
I had been told, resided a Malik particularly well-acquainted
with the Mahában region. I found in him a very intelligent
old man ready to describe what he had seen on frequent visits
to his Amazai friends, who are in the habit of grazing their cattle
on the mountain. He knew well the ruins of *Sháhkót*. He
described them as situated on a rocky spur near the highest point of
Mahában and to the north-east of it. Both the village of *Malka*
(once the seat of the Hindústáni fanatics and burned after the
Ambéla Campaign, 1863) and the Indus could be seen from the
plateau occupied by the ruins. I was particularly glad to note in
the course of my examination that the Malik's description of the
ruined fort agreed closely with the account given by Major Deane's
informant. The substantial accuracy of the latter account can
hence not be doubted. The ruins appear now to be overgrown
by dense jungle. The slopes of the mountain below Sháhkót
were described as steep and rocky on all sides, and particularly
so towards the Indus, where the ascent is by a narrow
path.

My informant did not stop at describing to me the mountain
of my desire, but also promptly offered, when alone with me and
my surveyor, to conduct me to it in person. Twelve hours' marching
and climbing might have sufficed to reach it. Under other circumstances the temptation would have proved too much for me. But
the thought of my escort and the promise I had given of rejoining
the troops before they cleared the pass left me no chance but
reluctantly to refuse this offer.

I then continued my ride to the large village of *Kuria* not
far off, which had been indicated to me as the extreme point
reached by a previous reconnaissance of the force. The village
lies on an alluvial plateau in the centre of the valley and opposite
to a bold fir-clothed spur which descends from the high *Sarpati*
Range, the continuation of Mahában to the west. From the rising

ground to the east of the village an extensive view opened down
the valley towards the Amazai territory and up to the snow-covered
heights of Mahában, comparatively so near and yet beyond reach.
I had but little time to enjoy it. The advanced hour and the
thought of the long ride yet before us necessitated an early return.
The road I followed back to Ambéla lay more to the north side
of the valley, but did not bring into view any further object of
antiquarian interest.

When Ambéla was reached in afternoon after a ride of about
9 miles from Kuria we found the large camp already deserted.
I followed the route taken by the troops into the wooded gorge
which leads to the *Ambéla Pass*, and overtook the rear guard of the
force close to the saddle of that famous defile, ever ' memorable
in the annals of frontier wars since the fights of 1863.

Ambéla Pass.

Rugged heights to the right and left crown the Kótal, which
Pathán tradition calls so forcibly *Qatalgarh*, ' the house of
slaughter.'* On them there were yet clearly visible rough stone
walls among the rocks, marking the sites of the " Eagle's Nest,"
the " Crag Picket " and other positions which were held so heroically
and at the cost of so much blood during those weeks of a desperate
struggle. I had thus the satisfaction of casting my farewell look
towards Bunér as one of the last who left its soil, and from a
spot full of historical associations, not less stirring because they
were modern. I derived some consolation from the memories of
that other Bunér campaign. From the point of view of antiquarian
research I had reason to regret the short duration of the present
expedition. Yet it was evident that its almost too rapid success
had its compensations in another direction.

There was little to remind me of those days of hard fighting
as I passed through the long winding ravines full of a luxuriant

---

* I cannot refrain here from drawing attention to the series of splendid
ballads in which Afghán popular poetry commemorates the events that played at
this site. My lamented friend J. Darmesteter has reproduced them, with a
masterly translation, in his *Chants populaires des Afghans*.

vegetation down to the southern foot of the pass. Apart from
the long files of ammunition mules passed on the way there were
only a few buffaloes, captured as a last lucky prize by a rear guard
picket on the heights near the pass, to show that we were
leaving an enemy's country. It was dark when I reached Surkhábi
at the mouth of the pass and in British territory, and night before
I arrived at the camp pitched near the little town of Rustam. Thus
a long day of nearly forty miles' ride and march brought my tour
with the Bunér Field Force to a close.

Bakhsháli.   On the following day I rode into Mardán, visiting on the way
a few old sites close to Rustam and near *Bakhshálí*. Those near
the former place have already been referred to by General Cunningham in his Archæological Survey Reports. At the latter place
I enquired particularly after the find-spot of the interesting ancient
birch-bark manuscript which was discovered here 17 years ago
and has since been edited by Dr. Hoernle.* I had the chance of
discovering the village chaukídár who had actually been the finder,
and was taken by him to the exact spot where the manuscript
was unearthed. As I think the site has not been accurately
indicated before its brief description may be useful.

The spot is at the north-west end of a series of ancient
mounds known as *Pandhérei*. They stretch in the direction
from north-west to south-east and for a length of about half
a mile to close the south-west corner of the present village.
The mounds rise to about 20 feet above the present ground level,
and are constantly dug into for the sake of building materials.
Walls of uncarved stone are found in many places at a depth of
3—8 feet from the present surface. Close to the spot where the
find was made a well had been sunk at the time and the field
near its east side dug down by 3 or 4 feet in order to bring it
more easily under irrigation. In the bank thus formed in the
mound to the east of the field the manuscript had come to light.
According to the account of the discoverer it was only 2—3 feet

---

* See *Indian Antiquary*, XVII, pages 33 *sqq.*

below the present surface placed between two stones and embedded in earth. As there are no visible traces of walls near the spot it may be, assumed that the manuscript was originally removed from some other place and buried here in the ground for protection or some other purpose. It may be added that there are numerous ancient wells near the Pandhérei site. One of them which is close to the north of the central mound has been recently cleared. It is circular and shows courses of solid ancient masonry, exactly of the same type as seen in the old well near the Sunigrám Stúpa. According to my informants more of these ancient wells in the neighbourhood would be cleared if experience did not show that they do not draw water or soon run dry. Does this observation indicate a change in the level of the subsoil water?

Arrived at Mardán, where the General Blood's Division broke up, I was engaged during the next few days in revising my materials and arranging for the preparation of the drawings attached to this report. I subsequently proceeded on a brief visit to Malakand in order to communicate personally to Major Deane the main results of my Bunér tour which he had done so much to facilitate. After another short stay at Mardán spent in preparing the preliminary portion of this report I returned to Lahore, where I resumed charge of my office on the 1st February 1898.

II.—Notes on the Ancient Topography of Buner.

Having completed my account of the ancient remains surveyed in Bunér I shall proceed to examine briefly the results that may be derived from the materials now collected for the elucidation of the ancient topography of that region. It has appeared to me more appropriate to discuss these results together and in a separate chapter. For it is only by comparing the whole of the ancient notices we possess of Bunér with the archæological data now available that we can arrive at approximately safe conclusions regarding the identification of several ancient sites.

The ancient notices of Bunér I allude to can unfortunately at present not be found in the form of inscriptions or in Indian historical records. Nor can they be gathered from the accounts which have reached us of Alexander's exploits in these regions. In view of what has been said above as to the probable identity of Mount Mahában with Alexander's Aornos, it appears possible that the great invader actually passed through a part of Bunér on his way from the valleys of the Panjkora and Swat. But the references by his historians to localities in this direction (*Ora, Basira, Dyrta*) are so vague and partly contradictory that guesses as to their identification can in the present state of our knowledge scarcely answer any useful purpose.[*]

Chinese Notices.

We are indebted for those notices exclusively to the narratives of the Chinese pilgrims who either on their way to Gandhára or in pious excursions from the latter had occasion to visit the sacred Buddhist sites in Udyána.

That the present territory of Bunér must have been comprised in the ancient Udyána had been recognized long ago by Sir Alexander Cunningham and V. de St. Martin when they endeavoured to map out the corresponding portions of the pilgrims'

---

[*] For a convenient summary of such guesses regarding places connected with Alexander's march towards Aornos, compare Dr. M'Crindle's *Invasion of India by Alexander the Great*, pages 72 *sqq.*, 335 *sqq.*

travels.*   But as long as the Swat Valley and the mountain
territories bordering on it remained wholly inaccessible to
Europeans and hence to a great extent a *terra incognita* also
from a geographical point of view, the elucidation of details
affecting the ancient topography of any one of these regions was
manifestly impossible.   Even now, when the veil has been partially
lifted, the task could scarcely be attempted with any hope of
success, were it not for the fortunate circumstance which supplies
us in the site of the ancient capital of Udyána with a fixed and
safe starting point for our enquiry.

I refer to the identification of the town of *Mangali* (*Mung-*   Position
kie-li) which HIUEN TSIANG, the latest and most accurate of those of Mangali.
pilgrims, mentions as the residence of the kings of Udyána.†
This is undoubtedly the present *Manglaur* in Upper Swat, which
is still remembered in local tradition as the ancient capital of the
country.   This identification was first proposed by V. DE
ST. MARTIN.   It has since been confirmed beyond all doubt by the
examination of the extant remains both at Manglaur and lower
down in the Swat Valley.‡   It has a special importance owing to the
fact that Hiuen Tsiang and also the earlier pilgrim SUNG-YUN
(A. D. 520) take the royal city as their starting point in giving
the direction and distances for the various sacred sites described
by them in Udyána.   Taking into account the ascertained position
of *Manglaur* at the point where the spurs descending to the north

---

* See *Cunningham, Ancient Geography of India*, pages 61 sq. ; *V. de St.
Martin, Mémoire Analytique sur la carte de l'Asie Centrale and de l'Inde*,
pages 313 *sqq.*

† See *Si-yu-ki*, transl. Beal, I., page 121.

See *Mémoire Analytique*, page 314, where the correct derivation of *Manglaur* (Manglavor) from Skr. *Mangalapura* is also indicated.   Hiuen Tsiang's
Moung-kie-li (to be read *Mangali*, see *St. Julien, Méthode pour déchiffrer les
noms sanscrits*, page 156) represents a shorter form *Mangala*, abbreviated
bhimavat, like *U-to-kia-han-cha* (*i.e.,** Udakahánda) for *Udabhándapura*, the ancient name of Waihand-Und on the Indus.

‡ See Major H. A. DEANE's paper " *Note on Udyána* and *Gandhára* " in the
Journal of the Asiatic Society, London, 1896, page 656.   Major Deane during
the reconnaissance made into Upper Swat in August last after the siege of
Malakand was able to pay a flying visit to the neighbourhood of Manglaur,
which abounds in ancient remains.   He there was able to recognize several of the
Stúpas mentioned by Hiuen Tsiang.

from Mount Dosirri meet the Swat River and turn it to the west (circ. 72° 28', long. 34° 48' lat.), it is clear that we must look for the ancient sites of Bunér among those localities of Udyána which the pilgrims describe as situated to the *south* of Mangali.

Hiuen
Tsiang's
account. The fullest account we receive of these localities is that preserved in the *Si-yu-ki* or "Records of the Western Countries" of Hiuen Tsiang, who visited Udyána from Udabhánda or Und on the Indus towards the close of 630 A. D. *

We leave aside for the present the reference made in his narrative to Mount *Hi-lo*. It is described as situated 400 *li*, or approximately 66 miles to the south of *Mung-kie-li*, and in view of this great distance cannot have been situated in Bunér proper. We are then first taken to the *Mahávana* convent. It lay about 200 *li* south from the capital by the side of a great mountain. The legend connected with it represented Buddha to have practised here in old days the life of a Bodhisattva under the name of *Sarvadarája*. Seeking a refuge from his enemy he had abandoned his kingdom and come to this place. There he met a poor Brahman who asked for alms. Having nothing to give him owing to his own destitute condition, Buddha had asked to be bound as a prisoner and to be delivered to the king, his enemy, in order that the Brahman might benefit by the reward given for him.

"To the north-west of the *Mahávana* Sangháráma one descends from the mountain and after proceeding for 30 or 40 *li* arrives at the *Mo-su* Sangháráma."† At this site the name of which is explained by the Chinese editor to mean 'lentils' and must hence probably be restored into *Mo-su-lo* (Skr. *masúra*), there was a Stúpa about 100 feet in height, and by the side of the latter a great square stone which bore the impress of Buddha's foot. When

---

* See *Si-yu-ki*, transl. Beal, I., pages 123 *sqq.*

† I have followed in the above abstracts Beal's translation, modifying its expressions only in a few places where the French version of Stan. Julien appeared to supply a more precise wording.

Buddha in old time p'anted his foot at this spot, "he scattered a
Koti of rays of light which lit up the Mahávana Sangháráma, and
then for the sake of Devas and men he recited the stories of his
former births. At the foot of this Stúpa is a stone of yellow-white
colour which is always damp with an unctuous moisture. This is
where Buddha, when he was in old time practising the life of a
Bodhisattva, having heard the true law, broke one of his bones
and wrote [with the marrow] sacred books."

Going west 60 or 70 *li* from the Mo-su convent Hiuen Tsiang
notices a Stúpa built by King Asoka. Here was localized the
well-known legend which records how Tathágata, when practising
the life of a Bodhisattva as Rája *Sibika*, had cut his body to pieces
to redeem a dove from the power of a hawk.

The short distances which Hiuen Tsiang indicates between
these three sacred sites show clearly that they must all have been
situated somewhere within Bunér territory. And in full agreement
with this conclusion we find that the two earlier pilgrims, FA-HIEN
and SUNG-YUN, who do not know the Mahávana Sangháráma, but
mention the other two sites of Hiuen Tsiang's account, also place
the latter distinctly to the south of the royal city of Udyána, *i.e.*, in
Bunér.

Fa-hien's notice

Fa-hien* who had arrived in ' *Wu-chang* ' (Udyána) about
403 A.D., and had spent the summer retreat there, descended
thence south and arrived in the country of *Su-ho-to*, where Buddhism
was flourishing. There was in it the place where in a former
birth "the Bodhisattva cut off a piece of his own flesh and with it
ransomed the dove .... On the spot the people of the country
reared a tope adorned with layers of gold and silver plates."
"The travellers, going downwards from this towards the east, in
five days came to the country of Gandhára." It cannot be doubted
that the Stúpa seen by FA-HIEN was that connected with

---

* See *Record of Buddhist Kingdoms*, translated by J. Legge, 1886, pages 29 *sqq.*

the legend of S.bikarája, which Hiuen Tsiang mentions a short
way to the west of the Mo-su convent. It is equally evident
that the district of *Su-ho-to*, in which it lay, must be identified
with the present Bunér. Arguing from the position indicated for
Su-ho-to by its mention to the south of Udyána and on the way to
Gandhára, General Cunningham* had already rightly recognized
that the territory thus designated could not have been the large
valley of the Swat River itself, as others have assumed, but that
the name must have been limited to the smaller valley of Bunér.

Sung-Yun's
account.

Evidence equally convincing as that just discussed may be
drawn from *Sung-Yun's* narrative. Sung-Yun, who visited the
' U-chang country ' towards the close of A. D. 519 as an imperial
envoy, notices to the south of its royal city the place where Buddha
in a former age "peeled off his skin for the purpose of writing upon it, and broke off a bone of his body for the purpose of
writing with it. Asoka Rája raised a pagoda on this spot for the
purpose of enclosing these sacred relics. It is about ten *chang*
(120 feet) high. On the spot where he broke off his bone the
marrow ran out and covered the surface of a rock which yet re-
tains the colour of it, and is unctuous, as though it had only recently
been done."† The place is spoken of by Sung-Yun as situated in
the ' *Mo-hiu* ' country.‡ Though we are unable to account for
this name, the description shows clearly that the Stúpa here referred to can be no other but the one mentioned by Hiuen Tsiang in
connection with the Mo-su Sangháráma.

In view of this identity of the site it is of interest to compare
the different indications given by the two pilgrims as to its position. Whereas Hiuen Tsiang places the Mo-su Sangháráma 30

---

* Compare *Ancient Geography*, page 82.

† See *Si-yu-ki*, transl. Beal, Introduction, page xcvii ; compare also the translation given by A. Remusat from an extract in the *Pian-Tian, Foeh-houe-ki*, page 50.

‡ *Mo-hiu* is possibly only another attempt to reproduce in Chinese characters the
local name which is given as *Mo-su* in Hiuen Tsiang's narrative. It should be noted
that the text of Sung-Yun's report seems in a far less satisfactory condition,
especially in regard to names, than that of Fa-hien or of the Si-yu-ki ; compare
BEAL's *Introduction* to the latter, page xcvii, note 68.

or 40 *li* to the north-west of the Mahávana monastery and the latter again about 200 *li* south of *Mung-kie-li*, Sung-Yun, who also starts from the royal city of Udyána, puts the former site at a distance of "more than 100 li" to the south of it. Apart from the identity of the bearings the two statements agree also close enough in respect of the distances. It must be remembered that the expressions of the texts distinctly indicate approximate measurements; allowance must further be made for the different length of the several routes which the pilgrims might have chosen for their journey from Upper Swat into Bunér.

The records of the Chinese travellers have shown us that among the sites of antiquarian interest described by them in or near Udyána there are three for the identification of which we have to look within the limits of modern Bunér. From a comparison of these accounts we have seen that the data they furnish regarding these sites are consistent among themselves, and hence evidently accurate. As information has now become available also as regards the actual topography of Bunér and the most prominent of its ancient remains, an attempt may well be made to trace the sites of those Stúpas and monasteries among the extant ruins of the territory.

The task thus set to us might be looked upon as partially solved or at least greatly facilitated, if the suggestion thrown out by General CUNNINGHAM of Mount Mahában having taken its name from the *Mahávana* monastery of Hiuen Tsiang could be accepted as probable.* This, however, is not the case. However tempting the similarity of the two names is upon which General Cunningham's conjecture was solely based, yet it is easy to show that this location meets with fatal objections both in the bearing and the distance indicated for the site in Hiuen Tsiang's narrative. The latter speaks of the *Mahávana* Sangháráma as

Mahávana Vihára.

---

* See *Archæological Survey Reports*, II, page 98; *Ancient Geography*, page 92.

situated 200 *li* to the south of *Mung-kie-li*. In reality Mount
Mahában lies to the *south-east* of Manglaur, as can easily be ascertained from the relative position shown on the accessible maps
for the trigonometrically fixed peaks of Dosirri and Mahában.*
In the same way it can be shown that the measurement of 200 *li*
does by no means agree with the actual distance by road between the two places.

Hiuen
Tsiang's
road measure-
ments.

In judging of this point it must be remembered that the distances between two places as recorded by the Chinese pilgrims
can have been derived only from approximate estimates of the
length of road traversed by them or their informants. They must
hence in a mountainous country be invariably much in excess of
the direct distances as measured on a modern survey map. The
examination of numerous cases, in which distances between well-
known localities have thus been recorded in road-measure, shows
that these measurements exceed by at least one-fourth, and in difficult country more nearly by one-third, the direct distance calculat-
ed on the maps.†

Keeping this in view it will be easy to recognise that Hiuen
Tsiang's Mahávana monastery cannot be looked for so far away as
Mount Mahában. The direct distance between the trigonometrically fixed peak of Mount Mahában and the position which the
field survey carried into Upper Swat during the operations of
last August ascertained for Manglaur, is exactly 40 miles, measured
on the map "as the crow flies." If we make to this distance the
above explained addition of one-fourth, which in view of the natural
obstacles of the route—the high range between Swat and Bunér
and the second hill range between the latter and the Chamla
Valley—must appear very moderate, we obtain a total distance by

---

* See Map " *District of Peshawar*," published by the Survey of India Office, 1884,
4 miles to 1 inch.

† See V. DE ST. MARTIN, *Mémoire Analytique*, page 259. Compare also
CUNNINGHAM, *Ancient Geography of India*, page 48.

road of not less than 50 miles. This minimum estimate of the
real road distance, when converted into Hiuen Tsiang's *li* at the
value of one-sixth of a mile for the *li*, as deducted by General
Cunningham from a series of careful computations,* gives us *three*
hundred *li* against the *two* hundred *li* actually recorded in the
pilgrim's narrative.

The difficulties in which the suggested identification of Hiuen
Tsiang's monastery with Mount Mahában would involve us become
still more prominent if we compare this location with another of
Hiuen Tsiang's topographical data bearing on Udyána and one
more easy to verify. I mean the statement made at the close of
Book II of the *Si-yu-ki*. There we are told that the pilgrim proceeding to the north from *U-to-kia-han-cha*, passed over some
mountains, crossed a river, and after travelling 600 *li* or so arrived
at the kingdom of *U-chang-na* or Udyána.† U-to-kia-han-cha is
undoubtedly the present *Und* on the Indus, the ancient capital of
Gandhára.‡

From the analogy of numerous passages in Hiuen Tsiang's
narrative, where the distances to capitals of neighbouring territories are indicated in a similar fashion, it is clear that the distance
here given to 'the kingdom of U-chang-na' must be understood
as referring to the capital of this territory, *i.e.*, *Mung-kie-li* or
Manglaur. Referring now to the relative position of Und and
Manglaur as fixed by modern surveys, we find that the capital of
ancient Udyána lies almost exactly due north of Und and at a direct
distance of 57 miles as measured on the map.

We do not receive any distinct information as to the route
which Hiuen Tsiang actually followed. Yet from the correct

---

* Compare *Ancient Geography*, page 571.

† See *Si-yu-ki*, transl. Beal, I, page 118. By the river here mentioned the
Barandu must be meant. But it should be noted that in Stan. Julien's translation
the word corresponding to 'river' is rendered by 'des vallées.'

‡ Compare REINAUD, *Mémoire sur l'Inde*, page 156, and my *notes on the history
of the Sáhis of Kabul*, page 7.

indication of the direction to the north and on general grounds it may
safely be assumed that he proceeded by one of the direct routes
leading through Bunér. The increased length of Hiuen Tsiang's
road measurement, 600 *li*, against the direct distance on the map, is
in the light of the explanations given above easily accounted for by
the natural difficulties of the track. These could not have been
appreciably smaller on the journey from Manglaur to Mahában,
which leads practically through the same mountain region. How
then, if the proposed identification of the Mahávana Sangháráma
with Mount Mahában is maintained, are we to understand the
great disproportion in the recorded distances,—the 200 *li* of one
journey against the 600 *li* of the other, where the direct distances
from point to point are 40 and 57 miles respectively?

Mahávana :
Panjkôtai.   It is evident from these considerations that the location of the
Mahávana monastery on Mount Mahában, based solely on a coinci-
dence of names, cannot be maintained. There remain thus for our
guidance only the facts of the actual topography of Bunér and that
knowledge of its extant ruins which the tour described in this report
has furnished. Reviewing then the most prominent of the ancient
sites surveyed we can scarcely fail to note the remarkable agreement which the ruins of *Panjkôtai* (Sunigrám), *Gumbatai* (Tursak)
and *Girárai* present with the three sacred spots specified in the
Chinese accounts both as regards their character and their relative
position.

We start from Manglaur as our fixed point. Referring to the
latest survey we find that Sunigrám lies almost due south of it,
exactly in the position indicated for the Mahávana monastery. The
nearest route between the two places lies over the Khalfl Pass (west
of Dosirri) and then *viâ* Gókand down to Pádsháh and Elai. It
measures on the map about 26 miles, which distance converted
according to the value previously indicated corresponds to about
156 *li*. If on the basis of the explanations already given, we add

to this distance on the map one-fourth in order to obtain the approximate road measurement, we arrive at the result of 192 *li*. This agrees as closely as we can reasonably expect with the 200 *li* of Hiuen Tsiang's estimate.

The pilgrim's description of the Mahávana monastery as situated "by the side of a great mountain" is fully applicable to the Panjkótai ruins. Even the absence of any reference to a Stúpa in connection with this monastery acquires significance in view of the fact that among the ruins, as described above, we fail to trace the remains of a Stúpa of any size.

The next stage of Hiuen Tsiang's itinerary to the *Mo-su* monastery takes us down the mountain to the north-west of the Mahávana Sangháráma for a distance of 30 or 40 *li*. Here the correspondence is again most striking. It is exactly to the north-west of the Panjkótai ruins, and after descending from the steep hill side on which they are situated, that we reach the *Gumbatai* site near Tursak. Its actual distance by road is about 6 miles, which corresponds to 36 *li*, or the mean of the approximate figures given by the pilgrim. Here we have no difficulty in recognizing the high Stúpa mentioned both by Hiuen Tsiang and Sung-Yun in the still extant mound, which even in its ruined condition forms a striking feature of the site. It can scarcely surprise us that the rapid survey of the ruins failed to bring to light here the stone at the foot of the Stúpa which according to the pious tradition marked the spot where Buddha had broken a bone of his body to write sacred texts with his marrow. The description of the site given above shows to what depth the base of the Stúpa is now hidden under débris.

Mo-su: Gumbatal.

Going 60 or 70 *li* to the west of the Mo-su Vihára, Hiuen Tsiang had visited the Stúpa reared over the spot where Buddha, according to the pious legend noticed also by Fa-hien, had sacrificed his body to ransom the dove. The bearing and distance here indicated agree so accurately with those of the ruined mounds near Girárai relative to Gumbatai that I do not hesitate to propose the

Girárai: Stúpa of 'Doveransoming.'

identification of the former with the sacred site referred to by the
two pilgrims. The ruined Stúpas of Alí Khán Kóte lie as above
indicated about 1½ miles to the west of Girárai village. The
distance from the latter place to Tursak on the direct track I
marched by was estimated by me at the time at about 7 miles.
The Gumbatai site again is, as already stated (page 24) 1½ miles
distant from Tursak. The total of these measurements is 10 miles,
which represents exactly the 60 *li* of Hiuen Tsiang's estimate.
There is the same accurate agreement as regards the direction,
the map and my own notes showing Girárai to be situated
almostly exactly due west of Tursak.

Route to Gaudhára.

There are two observations contained in the accounts of the
Chinese pilgrims which enable us to test at this point our chain of
identifications. Fa-hien's narrative (see above, page 55) tells us
that the travellers going downwards from the spot where Buddha
ransomed the dove, towards the east, in five days came to the
country of Gandhára. From the remarks which follow, it can be
concluded with great probability that the road distance here given
by Fa-hien was measured to the spot 'where Buddha in a former
birth had given his eyes in charity for the sake of a man,' and
where a great Stúpa had been erected in honor of this legendary
event. It is to be regretted that the sacred site here meant cannot
yet be identified. Sung-Yun also mentions it; but from his somewhat
confused account it can only be gathered that it lay somewhere in
the central part of the Yuzufzai plain.* A similar conclusion can
be drawn also from Fa-hien's own statement, who speaks of having
reached *Chu-cha-shi-lo*, or the place of 'the head offering,' the
well known site of Taxila, after a seven days' march to the east of
Gandhára, *i.e.*, of the spot already specified.†

On the first look it might appear strange that Fa-hien in order
to go from the Girárai site to the central part of Gandhára or

*See *Si-yu-ki*, transl. by Beal, p. clii.

†*Si-yu-ki*, p. xxxii. Taxila, marked by the ruins of the present Sháh-ki Dheri,
is placed by all Chinese accounts three marches to the east of the Indus; see
Canningham, *Ancient Geography*, page 104.

Yusufzai should proceed in an easterly direction, and should take five days to accomplish the journey. A reference to the map and a consideration of the ordinary routes still followed to the present day will, however, easily explain this.

Leaving the sacred site of the 'dove-ransoming' Fa-hien may naturally be supposed to have taken the most convenient and frequented route. In view of the topographical features of the country this would have been in his days just as now the route which leads first to the east down the Barandu Valley and then crosses the range of hills by the Ambéla Pass down to Rustam, an important site already in ancient times.* It is practically this route which was followed by the late expedition. On it five daily marches of the customary length would still be counted for the journey from Giráraí to Mardán, which latter place in view of its central position may here be taken as an approximate substitute for the site of 'the eye-offering.'†

A second test for the correctness of our proposed identifications is supplied by a statement of Hiuen Tsiang. He informs us that " going north-west from the place where Buddha redeemed the dove, 200 *li* or so, we enter the valley of *Shan-ni-lo-shi* and there reach the convent of Sa-pao-sha-ti."‡ Major DEANE in his very instructive " Note on Udyána and Gandhára " has proposed to identify the *Shan-ni-lo-shi* of the records with the large Adinzai Valley, which opens to the north of the Swat River near the present Fort Chakdarra. § The careful examination I was able to make during my two tours in the Swat Valley of the several topographical and archæological facts bearing on this question has convinced me that Major Deane has in this, as in other instances, been guided by the

Route to Shan-ni-lo-shi

Ancient Geography, page 65.

The probable stages would be Karapa or Sunigrám; Ambéla; Rustam; Bakhshali—all places which either by their remains or position can lay claim to importance from early times.

‡See *Si-yu-ki*, transl. Beal, i., page 125; *Mémoires de H. Th.*, i., page 137.

§Compare *Journal of the Royal Asiatic Society of Great Britain*, 1896, page 657.

right antiquarian instinct. I hope to discuss this point in a separate report on the remains of the lower Swat Valley. Here it may suffice to state that the *Sa-pao-sha-ti* convent with its high Stúpa must in all probability, as already recognized by Major Deane, be looked for among the several great ruined mounds which are found in the very centre of the valley close to the point where the present military road turns sharply to the west towards the Katgala Pass.

The general direction of the Adinzai Valley from Girárai is north-west, exactly as stated by Hiuen Tsiang. The nearest and apparently easiest route between the two places leads over the Banjir Pass down to the Swat River. Thence the road lies along the left bank of the latter to Chakdarra, which owing to its natural position must have at all times been the favourite point for crossing. Measured along this route the total distance on the map from Girárai to the central point of the Adinzai Valley above indicated amounts to 25 miles. This is almost exactly the distance which we have found above as the equivalent on the map of Hiuen Tsiang's 200 *li* between Manglaur-Mangali and Panjkótai-Mahávana. It is thus evident that, given the identical base of conversion, the 200 *li* of the pilgrim represent here with equal closeness the actual road distance between Girárai and Adinzai.

It is clear that we gain important evidence in favour of our chain of identifications in Bunér by being able to link also its western end with an ancient site of certain identity. The positions we have been led to assign to the Mahávana convent and the Stúpa of the ' dove-ransoming ' can thus each be independently tested by the bearings and distances recorded to known outside points. The positions hence mutually support each other.

We have made here the attempt to interpret the extant notices of ancient Bunér by means of the now available materials. It might be urged against it that these materials are still too scanty to permit of safe conclusions, and that in particular the

rapidity with which the survey of antiquarian remains had
to be effected on this occasion, was not likely to bring to notice
all important sites deserving consideration. In order to allay such
doubts it may be useful in conclusion to refer to an earlier
record. It shows that however hurried to my regret the examination of the territory has been, yet no important remains
above ground which were within reach, are likely to have wholly
escaped observation.

I refer to the curious information collected regarding Bunér
and the neighbouring regions by General A. COURT, one of
the French Officers in Ranjít Singh's service. It is contained in a
paper which was published by him in the Bengal Asiatic Society's Journal of 1839.* I did not see it until after my
return from Bunér. It contains, apart from purely geographical
notices regarding the mountain territories to the north of the
Peshāwar District, a series of conjectures as to the sites connected
with Alexander's carpaign in these regions, and what is far more
useful and interesting, a list of the ruins and in particular Stúpas
found in them. From the fullness of the latter notes and a
statement of General Court himself it is evident that they were
the result of careful and prolonged enquiries carried on through
native agents during the time that he was in charge of the Sikh
Forces in Peshawar. General Court had already before that time
testified his interest in antiquarian research by the systemtic
excavation of the Manikyála Stúpa and the valuable
numismatic materials he collected for Prinsep and other scholars.
We can, therefore, scarcely be surprised at the thoroughness
with which he had endeavoured in this instance to collect all
information obtainable from native sources regarding the extant
monuments of those territories.

If we compare the entries in his lists of " ruined cities " and
" of cupolas " † as far as they relate to Swat, with the ancient

General
Court's
notes on
Bunér.

* See *Collection of Facts which may be useful for the comprehension of Alexander
the Great's exploits on the Western Banks of the Indus*, by M. A. COURT, Ancient
Éléve de l'École Militaire de Saint-Cyr, *J. A. S. B.,* 1839, pages 306 *sqq.*

† See pages 307 *sq* and 311, *loc. cit.*

sites and buildings which have attracted pre-eminently our attention since that valley has been rendered accessible, we find almost all important remains still above ground duly noticed. The temple of Talásh, with its elaborate relievos, the Stúpas of Adinzai, the ruins of Baríkót, the great Stúpa of Shankardár, the mounds around Manglaur,—these and other striking remains find all due mention, though their names appear more than once strangely disguised in the General's spelling.

Having observed this laudable accuracy of the information recorded regarding Swat, I naturally turned with a good deal of curiosity to General Court's notes regarding Bunér. Might they not tell of ancient remains of evident importance which I had failed to notice? I was soon reassured on this score. I found that of the old sites named by General Court's informants in Bunér proper, all, with one doubtful exception, had actually been visited by me.

Notices of Stúpas.

Among the cupolas, * i. e. Stúpas, which are specially singled out for notice, we find " those of *Heniapoor*, one of which is near the village of *Fooraseuk*, and the other under Mount Jaffer. " It requires no great amount of philological acumen to recognize here in the General's (or his English translator's) ' *Fooraseuk* ' our Tursak, and in his ' *Heniapoor* ' the name of the village Anrapur, which we have noticed above as situated just opposite to the Gumbatai Stúpa. For the mistake in the first name the quasi-palæographical explanation (F misread for T) easily suggests itself. In the case of the second the peculiar Pushtu sound *nr* is evidently responsible for the deficient spelling.† It is clear that this notice refers in reality to one Stúpa, that of Gumbatai, which, as we have seen, lies near Tursak at the foot of Mount Jaffer and opposite Anrapur. Whether the kind of

---

* The word ' cupola ' is evidently intended as a rendering of the term ' Gumbaz ' (dome) which is uniformly applied in these regions to all ruined Stúpas and dome-shaped buildings ; see above page 11.

† For the same reason the name appears in the maps metamorphosed into *Angapur*.

' diplography ' noticeable in General Court's description is due to his having recorded two separate accounts without noticing that they referred to the same structure, or to some other misunderstanding, cannot be decided now.

The cupola near ' *Sonigheran,* ' which is next mentioned, can be no other than the great ruined Stúpa south of *Sunigrám.* By another " in the village of *Fakttahind* " is clearly meant the Stúpa of Takhtaband. The same clerical error or misprint as in *Fooraseuk-Tursak* accounts for the change of the initial consonant in the local name. The reference to a Stúpa in ' Caboolgheram, ' *i. e.* Kabulgrám on the Indus, agrees with information supplied to me. But this locality, which can scarcely be included in Bunér, was, of course, beyond the limit of my explorations.

General Court's list mentions after the cupola near ' Sonigheran ' the two found among the ruins situated at the foot of Mount *Sukker* near the village of *Riga.* The name ' Riga ' stands here evidently for *Réga,* the home of the ' Mad Fakír ' and our camp from the 15th to the 16th January. But as, notwithstanding repeated enquiries and comparatively close inspection, I failed to trace any conspicuous remains in the immediate vicinity of that village, I feel induced to suspect that General Court's informant in reality intended a reference to the ruins of Panjkótai above Sunigrám. *Réga,* a large village, is a far better known place than the small hamlet of Sunigrám, and as the direct distance between the two is scarcely more than 1½ miles, the above-named ruins could equally well be described as situated near *Réga.* I cannot identify " Mount Sukker. " The name may possibly be that of the hill, on a spur of which the Panjkótai Vihára is built. That the high vaulted halls of the latter should be included under the head of " Cupolas " could not surprise. In the same way we find the ruined monastery of *Chárkótli,* situated in the gorge south of Batkhéla, Swat, which I visited in December 1897 without tracing near it any Stúpa remains, referred to under that designation in General Court's list (No. 6, ' *Charkotlia* ').

If we add that besides the above notices General Court's paper contains also a correct account of the Hindu Tírthas on Mount Ilm, it will be acknowledged that his agents had taken evident care to ascertain and to report all ancient sites in Bunér which were likely to attract attention.

This observation can only help to assure us as to the results of our own survey. We have seen that the latter, however hurried, has not failed to take us to every one of the sites which were known to General Court's informants, and this though at the time I was wholly unaware of this earlier record. We may hence conclude that the ruins described in this report include most, if not all, of the more important ancient sites of Bunér. We are thus justified in looking among them for the remains of those sacred buildings which in the records of the Chinese pilgrims receive special mention.

*   *   *

Conclusion.

In concluding the account of my tour in Bunér it is my pleasant duty to record my sense of gratitude for the manifold help enjoyed by me. In the first place my sincere thanks are due to the Punjab Government and its present head, the Hon'ble Sir W. MACKWORTH YOUNG, K.C.S.I., who readily sanctioned the proposal of my deputation and agreed to meet its cost. By thus rendering my tour possible the Punjab Government have given once more a proof of their desire to further the objects of Indian antiquarian research. This, I trust, will be appreciated all the more as the field to be explored lay on this occasion beyond the limits of the Province.

The above pages have shown how much assistance I derived from the kind interest which Major H. A. Deane, C.S.I., has taken in my tour. Students of the antiquities of the North-West Frontier region know the valuable discoveries due to Major Deane's zeal for archæological exploration and his readiness to facilitate all researches bearing on those territories.

It is an equal pleasure to me to record publicly my sense of the great obligations I owe to the Military and Political authorities of the Bunér Field Force. Major-General SIR BINDON BLOOD, K.C.B., Commanding the Division, not only agreed in the kindest manner to allow me to accompany the expedition, but also showed on many occasions his interest in my work and his desire to facilitate it by all means at his disposal. His staff as well as the Political Officers attached to the Force, Mr. BUNBURY, C.S., and Lieutenant DOWN of the Punjab Commission, were ever willing to give me all needful assistance.

I feel particularly grateful to Brigadier-General SIR W. MEIKLEJOHN, K.C.B., Commanding the 1st Brigade, and his staff for the free scope they allowed for my movements. Personally I doubt whether a civilian on a similar mission could ever have met with a kindlier reception than that which was accorded to me among the officers of the Bunér Field Force.

M. Fasl Ilâhi, Draftsman, Public Work Department, who was deputed to accompany me, rendered valuable services by making accurate surveys and plans of all the more important sites and ruins. I must specially commend him for the readiness with which he volunteered for the duty, and the careful and intelligent way in which he carried out his work, often under somewhat trying conditions. Nor ought I to omit a grateful reference to the excellent marching of the Afrídí escorts furnished to me by the XXth Regiment Punjab Infantry which enabled me to make full use of the limited time available for my excursions.

PLAN OF RUINED BUILDING
NEAR
KINGARGALAI

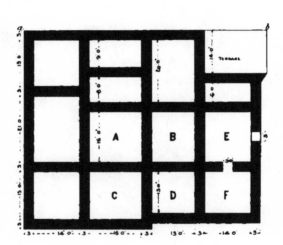

SCALE OF FEET

PLAN OF RUINED BUILDINGS

NEAR

KINGARGALAI

N

PLAN

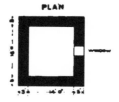

EAST SIDE ELEVATION

SCALE OF FEET

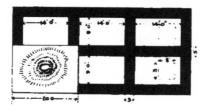

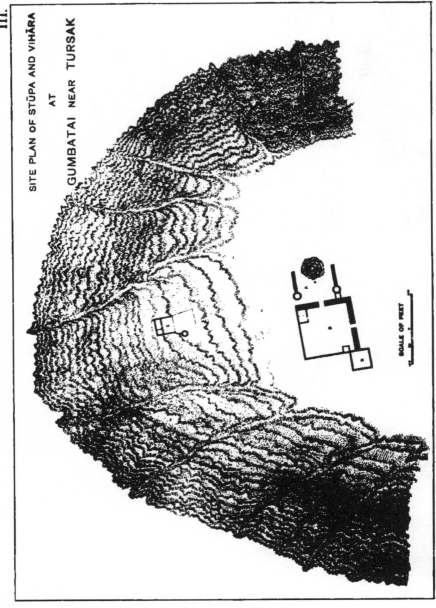

III.

SITE PLAN OF STŪPA AND VIHĀRA
AT
GUMBATAI NEAR TURSAK

SCALE OF FEET

IV

DETAIL PLAN OF STŪPA AND MONASTERY

AT

GUMBATAI NEAR TURSAK

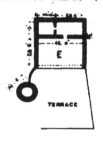

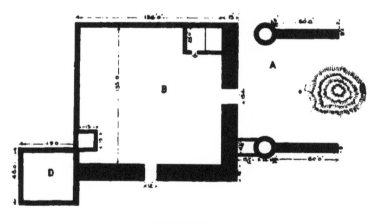

SCALE OF FEET

PLAN AND SECTION OF OLD WELL
AT
SUNIGRĀM

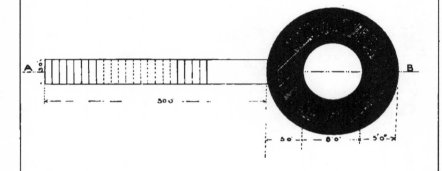

A 5'0" ———————— 30'0" ———————— B

5'0' — 8'0' — 5'0"

SECTION on LINE A.B.

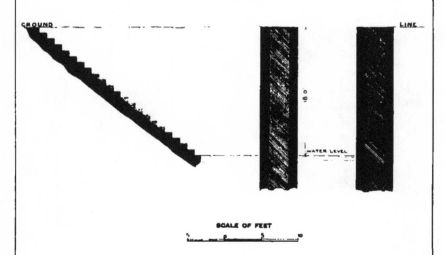

GROUND   LINE

18'0"

WATER LEVEL

SCALE OF FEET

5   0   5   10

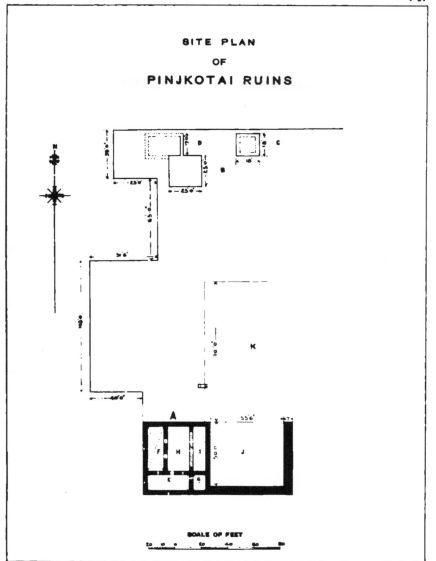

SITE PLAN
OF
PINJKOTAI RUINS

SCALE OF FEET